RELATIVITY THEORY: ITS ORIGINS AND IMPACT ON MODERN THOUGHT

MAJOR ISSUES IN WORLD HISTORY

Series Editor: C. Warren Hollister,
University of California, Santa Barbara

The Twelfth-Century Renaissance
C. Warren Hollister

The Impact of Absolutism in France:
National Experience under Richelieu, Mazarin, and Louis XIV
William F. Church

The Impact of the Norman Conquest
C. Warren Hollister

Relativity Theory:
Its Origins and Impact on Modern Thought
L. Pearce Williams

RELATIVITY THEORY: ITS ORIGINS AND IMPACT ON MODERN THOUGHT

Edited by

L. PEARCE WILLIAMS

John Wiley & Sons, Inc.

New York • London • Sydney • Toronto

Library of Congress Catalog Card Number: 68-30923
Cloth: SBN 471 94853 5 Paper: SBN 471 94854 3
Printed in the United States of America

TO DAVID

Series Preface

Traditionally, the reading program in a history survey course has usually consisted of a large two-volume textbook and perhaps a book of readings. This simple reading program requires few decisions and little imagination on the instructor's part and tends to encourage in the student the virtue of careful memorization. Such programs are by no means things of the past, but they certainly do not represent the wave of the future.

The reading program in survey courses at many colleges and universities today is far more complex. At the risk of over-simplification, and allowing for many exceptions and overlaps, it can be divided into four categories: (1) textbook, (2) original source readings, (3) specialized historical essays and interpretive studies, and (4) historical problems.

Having obtained an overview of the course subject matter (textbook), having sampled the original sources, and having been exposed to selective examples of excellent modern historical writing (historical essays), the student can turn to the crucial task of weighing various possible interpretations of major historical issues. It is at this point that memory gives way to creative critical thought. The "problems approach," in other words, is the intellectual climax to a thoughtfully conceived reading program and is indeed the most characteristic of all approaches to historical pedagogy among the newer generation of college and university teachers.

The historical problems books currently available are many and varied. Why add to this information explosion? Because the Wiley "Major Issues" series constitutes an endeavor to produce something new that will respond to pedagogical needs thus far unmet. To begin with, it is a series of individual volumes—one per problem. Many good teachers would much prefer to select their own historical issues rather than being tied to an inflexible sequence of issues imposed by a publisher and bound together between two covers. Second, the Wiley "Major Issues" series is based on the idea of approaching the significant problems of history through a deft interweaving of primary sources and secondary analysis, fused together by the skill of a scholar-editor. It is felt that the essence of an historical issue can be satisfactorily probed neither by placing a body of undigested source materials into the hands of inexperienced students nor by limiting these students to the controversial literature of modern scholars who debate the meaning of sources that the student never sees. The "Major Issues" series approaches historical problems by exposing students to the finest historical thinking on the issue and some of the evidence on which this thinking is based. This synthetic approach should prove far more fruitful than either the raw-source approach of the exclusively second-hand approach, for it combines the advantages—and avoids the serious disadvantages—of both.

Finally, the editors of the individual volumes in the "Major Issues" series have been chosen from among the ablest scholars in their fields. Rather than faceless referees, they are historians who know their issues from the inside and in most instances have themselves contributed significantly to the relevant scholarly literature. It has been the editorial policy of this series to permit the editor-scholars of the individual volumes the widest possible latitude both in formulating their topics and in organizing their materials. Their scholarly competence has been unquestioningly respected; they have been encouraged to approach the problems as they see fit. The titles and themes of the series volumes have been suggested in nearly every case by the scholar-editors themselves. The criteria have been (1) that the issue be of relevance to under-graduate lecture courses in history and (2) that it be an issue which the scholar-editor knows thoroughly and in which he has done creative work. In general, criterion (2) has been given precedence over criterion (1). In short, the question, what are the significant historical issues today, has been answered not by general editors or sales departments but by the scholar-teachers who are responsible for these volumes.

University of California, Santa Barbara C. WARREN HOLLISTER
JUNE 1968

Contents

RELATIVITY THEORY: ITS ORIGINS AND IMPACT ON MODERN THOUGHT

INTRODUCTION

There is one thread that runs through the history of physics in the nineteenth century. It is the story of the gradual expansion of the role of the ether in explaining physical phenomena. In the eighteenth century the ether was quietly nudged out of the center of physical speculation, even though the great Sir Isaac Newton had had kind words for it. In its place was substituted "action at a distance" by material particles endowed in some mysterious fashion with forces that literally "o'er leaped" space itself to influence other particles of a similar kind. Thus in the early 1800's one could describe a universe of seemingly unobjectionable simplicity. There were seven "elementary" particles:

1. Ponderable matter, whose basic force was canonized in the various weights of the individual atoms by John Dalton in 1808.

2. The corpuscles of light that were shot like bullets out of luminous bodies, bounced off reflectors like billiard balls, and were drawn into transparent media by a peculiar attraction that forced them to traverse these bodies with a speed greater than that at which they traveled through empty space.

3. The particles of heat, or caloric, which by their mutual repulsion caused the expansion of bodies with increasing temperature.

4 and 5. The two species of electrical particles, positive and negative. Two seemed necessary, for if there were only one kind whose particles repelled one another it was rather difficult to understand why the *lack* of the electrical fluid in two bodies whose ponderable atoms attracted one another should lead to the mutual repulsion of these *negatively* charged bodies. Thus there were particles of positive electricity and particles of negative electricity that acted upon one another according to precise laws laid down by Charles de Coulomb in the 1780's.

6 and 7. The same reasoning applied to magnetism. By analogy with electricity, boreal and austral magnetism could best be represented by particles of two kinds acting upon one another according to precisely determined experimental (Coulombian) laws.

Here were the makings for a complete physical theory of the world. It was clearly enunciated by the great French mathematical physicist Pierre Simon de Laplace. Particles, endowed with forces acting on one another at a distance through empty, Euclidean space, made up the universe. Except for light and heat, for which the precise form of the force laws was not yet known, these particles obeyed the Newtonian laws of attraction and repulsion.

The history of nineteenth century physics is, in large part, the story of the elimination of these "elementary" particles, with the possible exception of ponderable matter. Light was the first to go, with the enunciation of a wave theory by Thomas Young and Augustin Fresnel in the early 1800's. The triumph of the wave theory, in turn, brought about the revival of the eighteenth century ether, for it seemed only common sense to assume that if light were a wave, there must be some material substratum that did the waving; this was the luminiferous ether. Caloric soon followed suit. When radiant heat was found to act like light, it was, like light, assumed to be an ethereal vibration. Ordinary heat by the 1860's was considered by most physicists to be the kinetic energy of the particles of ponderable matter. Yet even here the ether intruded. Could not the atoms of ponderable matter be viewed as peculiar motions of the ether? Smoke rings, for example, were shown to have considerable stability, and there were some who felt that combinations of ether rings might be substituted for the ever growing number of chemical elements.

Electricity and magnetism by the 1870's could be treated with some confidence as strains and displacements of the ether. James Clerk Maxwell's famous equations expressed in succinct mathematical terms the various relations between electrostatic, electrodynamic, and electromagnetic forces, all of which could be visualized in terms of an all-pervading ether. By the end of the century Paul Drude, the editor of one of the most prestigious scientific journals of the day, could write a textbook, entitled *The Physics of the Ether*, that claimed to deal with most, if not all, of physics.

The elevation of the ether to a position of primacy in physics created certain anxieties that increased as time passed. Most physicists felt strongly and instinctively that they worked with the real world in their science. Yet, in spite of many ingenious attempts, the ether itself had never been isolated, weighed, smelled, seen, or tasted. Granted that it might be a "most subtile spirit," as Newton had described it, still a whole host of most subtile physicists ought to be able to devise *some* means of detecting it. One avenue seemed to be especially promising. In the early eighteenth century the English astronomer James Bradley had discovered a phenomenon known as the aberration of light, which, in terms of the wave theory of light, was interpreted to mean that the ether was stationary. The earth, then, in its annual revolution about the sun must move through this ether, creating an "ether wind," much as the movement of a convertible on a still summer day causes a "wind" for those riding in it. By the 1880's there were instruments available to measure this "wind," and the attempt was made. No "wind" could be detected! Was the experiment properly performed? Or were there errors that vitiated the results? The "wind," after all, was approximately of the intensity caused by a rather lethargic butterfly at five paces from the observer. There were those who felt disconsolate that there was no "wind," and therefore that physics had to face up to contradictory facts. Aberration "proved" that the ether was at rest; the Michelson-Morley experiment "proved" that the ether was in motion with the earth. There were those who were untroubled; the Michelson-Morley experiment had too many sources of error to be taken seriously. The ether *had* to exist or physics would be destroyed. There were also those, especially on the Continent, who had never heard of the Michelson-Morley experiment, for its fame, in many ways, was caused by the furor created by the Special Theory of Relativity and it is not safe to assume that its implications or even its facts were as widely known among physicists then as they were after 1905.

Here, then, lies one of the possible origins of the theory of relativity. The facts, by the end of the nineteenth century, spoke for themselves. The whole science of physics was based upon a theoretical entity that enjoyed the paradoxical property of being at rest and in motion—both at the same time! Something, clearly, had to give; the *facts* must force a reevaluation of physical theory. This is the conventional and most widespread view of the evolution of scientific theories. A theory is held until facts force its abandonment and the creation of a new theory. The facts precede the theory and regulate theorizing. There is no place, in this view, for imagination, speculation, aesthetics, or other individual, subjective factors. The facts pile up to build a theory; they continue to pile up until the theory collapses and another, larger pile takes its place.

The origins of the Special Theory of Relativity provide a particularly interesting example of the overthrow of one theory and the creation of a new one. To those who think that facts make theories, the historical path is a clear one; but the paradox of the motion of the ether is not the only possible source of discontent with the theories of classical physics. There is, too, a philosophical dimension that cuts deeply into the foundations of physical laws. The physicist may often rest content with laws of force acting between bodies; it is rare that he asks what, precisely, is meant by force, or mass, or inertia, or gravity; or, on a somewhat more abstract level, what kind of physical laws ought to be found in a universe created by a God who, at least in the tradition of Western civilization, is more often rational than not? Force, mass, inertia, and gravity, when looked at with a somewhat critical eye, lose their solidity and fade off into the smoke of semantics. The laws of nature, on the other hand, are solid and productive. But are they aesthetically satisfying? Do we feel, when viewing them, the same sense of balance that is felt in the contemplation of the Parthenon or Michelangelo's David? Are they, in fact, symmetrical and would the God known to Western civilization make an asymmetrical universe? The answer to the first question is that, in Maxwell's expression, the laws of nature are not symmetrical. Albert Einstein's answer to the second question would seem to be a resounding *No*. Was this, then, the *real* origin of the Special Theory of Relativity?

The publication of the Special Theory of Relativity in 1905 and the General Theory in 1916 created a stir throughout the Western world comparable only to that which greeted the publication of Charles Darwin's *Origin of Species*. The selections in the fourth part of this work speak for themselves and could be multiplied one hundredfold. Basically, however, they can be separated into two categories that, in a strange way, both reinforce and contradict one another. To the opponents of relativity the main point is that if relativity is true then common sense cannot cope with the physical world. Before 1905 it was possible for science to be explained to the layman in verbal terms which, although fuzzy, made sense. Henceforth this was impossible, for the special quality of the Special Theory was that it violated all the principles of common sense; or, to put it another way, science before 1905 could usually illustrate its most abstruse discoveries by means of a mechanical model; after 1905 the models were mathematical. The full sense of dismay felt by those trained in the classical tradition may be illustrated by Lord Kelvin's remark that he could understand something only if he could make a mechanical model of it.

The mathematization of nature, however, had its compensations, for there seemed to be a large subjective component in the new theories of relativity. Reality, Einstein seemed to be saying, depends on *your* viewpoint. Thus in art, psychology, sociology, and other fields the theory of relativity should free the individual from the tyranny of mechanical and mechanistic laws.

Hence the paradox. Relativity means the death of common sense, but the death, in all fields but physics, is relative.

PART I

The Origins of the Special Theory of Relativity

1 FROM *Sir Isaac Newton* *Mathematical Principles of Natural Philosophy*

In 1687 perhaps the most important book ever published in the history of science appeared in England. It bore the forbidding title Philosophiae naturalis principia mathematica (Mathematical Principles of Natural Philosophy). *Its author was the Lucasian Professor of Mathematics in Cambridge University, Isaac Newton.*

Newton's aim in the Principia *was to do just what the title suggested, to lay out those mathematical principles that, when applied to the problems of natural philosophy, might be expected to illuminate them. To do this Newton had first to set the mathematical stage. Thus at the very beginning he laid out his ideas on the nature of space and time, the fundamental arena in which physical events took place.*

SCHOLIUM

Hitherto I have laid down the definitions of such words as are less known, and explained the sense in which I would have them to be understood in the following discourse. I do not define time, space, place, and motion, as being well known to all. Only I must observe, that the common people conceive those quantities under no other notions but from the relation they bear to sensible objects. And thence arise certain prejudices, for the removing of which it will be convenient to distinguish them into absolute and relative, true and apparent, mathematical and common.

I. Absolute, true, and mathematical time, of itself, and from its own nature, flows equably without relation to anything ex-

SOURCE. Sir Isaac Newton, *Mathematical Principles of Natural Philosophy*, Andrew Motte, tr. (1729 edition), Florian Cajori, revised translation, Berkeley: University of California Press, 1947, pp. 6–12.

ternal, and by another name is called duration: relative, apparent, and common time, is some sensible and external (whether accurate or unequable) measure of duration by the means of motion, which is commonly used instead of true time; such as an hour, a day, a month, a year.

II. Absolute space, in its own nature, without relation to anything external, remains always similar and immovable. Relative space is some movable dimension or measure of the absolute spaces; which our senses determine by its position to bodies; and which is commonly taken for immovable space; such is the dimension of a subterraneous, an aerial, or celestial space, determined by its position in respect of the earth. Absolute and relative space are the same in figure and magnitude; but they do not remain always numerically the same. For if the earth, for instance, moves, a space of our air, which relatively and in respect of the earth remains always the same, will at one time be one part of the absolute space into which the air passes; at another time it will be another part of the same, and so, absolutely understood, will be continually changed.

III. Place is a part of space which a body takes up, and is according to the space, either absolute or relative. I say, a part of space; not the situation, nor the external surface of the body. For the places of equal solids are always equal; but their surfaces, by reason of their dissimilar figures, are often unequal. Positions properly have no quantity, nor are they so much the places themselves, as the properties of places. The motion of the whole is the same with the sum of the motions of the parts; that is, the translation of the whole, out of its place, is the same thing with the sum of the translations of the parts out of their places; and therefore the place of the whole is the same as the sum of the places of the parts, and for that reason, it is internal, and in the whole body.

IV. Absolute motion is the translation of a body from one absolute place into another; and relative motion, the translation from one relative place into another. Thus in a ship under sail, the relative place of a body is that part of the ship which the body possesses; or that part of the cavity which the body fills, and which therefore moves together with the ship: and relative rest is the continuance of the body in the same part of the ship, or of its cavity. But real, absolute rest, is the continuance of the body in the same part of that immovable space, in which the ship itself, its cavity, and all that it contains, is moved. Wherefore, if the earth is really at rest, the body, which relatively rests in the ship, will really and absolutely move with the same velocity which the ship has on the earth. But if the earth also moves, the true and absolute motion of the body will arise, partly from the true motion of the earth, in immovable space, partly from the relative motion of the ship on the earth; and if the body moves also relatively in the ship, its true motion will arise, partly from the true motion of the earth, in immovable space, and partly from the relative motions as well of the ship on the earth, as of the body in the ship; and from these relative motions will arise the relative motion of the body on the earth. As if that part of the earth, where the ship is, was truly moved towards the east, with a velocity of 10010 parts; while the ship itself, with a fresh gale, and full sails, is carried towards the west, with a velocity expressed by 10 of those parts; but a sailor walks in the ship towards the east, with 1 part of the said velocity; then the sailor will be moved truly in immovable space towards the east, with a velocity of 10001 parts, and relatively on the earth towards the west, with a velocity of 9 of those parts. . . .

As the order of the parts of time is immutable, so also is the order of the parts of space. Suppose those parts to be moved out of their places, and they will be moved (if the expression may be allowed) out of themselves. For times and spaces are, as it were, the places as well of themselves as of all other things. All things are placed in time as to order of succession; and in space as to order of situation. It is from their essence or nature that they are places; and that the primary places of things should be movable, is absurd. These are therefore the absolute places; and translations out of those places, are the only absolute motions.

But because the parts of space cannot be seen, or distinguished from one another by our senses, therefore in their stead we use sensible measures of them. For from the positions and distances of things from any body considered as immovable, we define all places; and then with respect to such places, we estimate all motions, considering bodies as transferred from some of those places into others. And so, instead of absolute places and motions, we use relative ones; and that without any inconvenience in common affairs; but in philosophical disquisitions, we ought to abtract from our senses, and consider things themselves, distinct from what are only sensible measures of them. For it may be that there is no body really at rest, to which the places and motions of others may be referred.

But we may distinguish rest and motion, absolute and relative, one from the other by their properties, causes, and effects. It is a property of rest, that bodies really at rest do rest in respect to one another. And therefore as it is possible, that in the remote regions of the fixed stars, or perhaps far beyond them, there may be some body absolutely at rest; but impossible to know, from the position of bodies to one another in our regions, whether any of these do keep the same position to that remote body, it follows that absolute rest cannot be determined from the position of bodies in our regions.

It is a property of motion, that the parts, which retain given positions to their wholes, do partake of the motions of those wholes. For all the parts of revolving bodies endeavor to recede from the axis of motion; and the impetus of bodies moving forwards arises from the joint impetus of all the parts. Therefore, if surrounding bodies are moved, those that are relatively at rest within them will partake of their motion. Upon which account, the true and absolute motion of a body cannot be determined by the translation of it from those which only seem to rest; for the external bodies ought not only to appear at rest, but to be really at rest. For otherwise, all included bodies, besides their translation from near the surrounding ones, partake likewise of their true motions; and though that translation were not made, they would not be really at rest, but only seem to be so. For the surrounding bodies stand in the like relation to the surrounded as the exterior part of a whole does to the interior, or as the shell does to the kernel; but if the shell moves, the kernel will also move, as being part of the whole, without any removal from near the shell.

A property, near akin to the preceding, is this, that if a place is moved, whatever is placed therein moves along with it; and therefore a body, which is moved from a place in motion, partakes also of the motion of its place. Upon which account, all motions, from places in motion, are no other than parts of entire and absolute motions; and every entire motion is composed of the motion of the body out of its first place, and the motion of this place out of its place; and so on, until we come to some immovable place, as in the before-mentioned example of the sailor. Wherefore, entire and absolute motions can be no otherwise determined than by immovable places; and for that reason I did before refer those absolute motions to immovable places, but relative ones to movable places. Now no other places are immovable but those that, from infinity to infinity, do all retain the same given position one to another; and upon this account must ever remain unmoved; and do thereby constitute immovable space.

The causes by which true and relative motions are distinguished, one from the other, are the forces impressed upon bodies to generate motion. True motion is neither generated nor altered, but by some force impressed upon the body moved; but relative motion may be generated or altered without any force impressed upon the body. For it is sufficient only to impress some force on other bodies with which the former is compared, that by their giving way, that relation may be changed, in which the relative rest or motion of this other body did consist. Again, true motion suffers always some change from any force impressed upon the moving body; but relative motion does not necessarily undergo any change by such forces. For if the same forces are likewise impressed on those other bodies, with which the comparison is made, that the relative position may be preserved, then that condition will be preserved in which the relative motion consists. And therefore any relative motion may be changed when the true motion remains unaltered, and the relative may be preserved when the true suffers some change. Thus, true motion by no means consists in such relations.

The effects which distinguish absolute from relative motion are, the forces of receding from the axis of circular motion. For there are no such forces in a circular motion purely relative, but in a true and absolute circular motion, they are greater or less, according to the quantity of the motion. If a vessel, hung by a long cord, is so often turned about that the cord is strongly twisted, then filled with water, and held at rest together with the water; thereupon, by the sudden action of another force, it is whirled about the contrary way, and while the cord is untwisting itself, the vessel continues for some time in this motion; the surface of the water will at first be plain, as before the vessel began to move; but after that, the vessel, by gradually communicating its motion to the water, will make it begin sensibly to revolve, and recede by little and little from the middle, and ascend to the sides of the vessel, forming itself into a concave figure (as I have experienced), and the swifter the motion becomes, the higher will the water rise, till at last, performing its revolutions in the same times with the vessel, it becomes relatively at rest in it. This ascent of the water shows its endeavor to recede from the axis of its motion; and the true and absolute circular motion of the water, which is here directly contrary to the relative, becomes known, and may be measured by this endeavor. At first, when the relative motion of the water in the vessel was greatest, it produced no endeavor to recede from the axis; the water showed no tendency to the circumference, nor any ascent towards the sides of the vessel, but remained of a plain surface, and therefore its true circular motion had not yet begun. But afterwards, when the relative motion of the water had decreased, the ascent thereof towards the sides of the vessel proved its endeavor to recede from the axis; and this endeavor showed the real circular motion of the water continually increasing, till it had acquired its greatest quantity, when the water rested relatively in the vessel. . . .

Wherefore relative quantities are not the quantities themselves, whose names they bear, but those sensible measures of them (either accurate or inaccurate), which are commonly used instead of the measured quantities themselves. And if the meaning of words is to be determined by their use, then by the names time, space, place, and motion, their [sensible] measures are properly to be understood; and the expression will be unusual, and purely mathematical, if the measured quantities themselves are meant. On this account, those violate the accuracy of language, which ought to be kept precise, who interpret these words for the measured quantities. Nor do those less defile the purity of mathematical and philosophical truths, who confound real quantities with their relations and sensible measures.

It is indeed a matter of great difficulty to discover, and effectually to distinguish, the true motions of particular bodies from the apparent; because the parts of that immovable space, in which those motions are performed, do by no means come under the observation of our senses. Yet the thing is not altogether desperate; for we have some arguments to guide us, partly from the apparent motions, which are the differences of the true motions; partly from the forces, which are the causes and effects of the true motions. For instance, if two globes, kept at a given distance one from the other by means of a cord that connects them, were revolved about their common centre of gravity, we might, from the tension of the cord, discover the endeavor of the globes to recede from the axis of their motion, and from thence we might compute the quantity of their circular motions. And then if any equal forces should be impressed at once on the alternate faces of the globes to augment or diminish their circular motions, from the increase or decrease of the tension of the cord, we might infer the increment or decrement of their motions; and thence would be found on what faces those forces ought to be impressed, that the motions of the globes might be most augmented; that is, we might discover their hindmost faces, or those which, in the circular motion, do follow. But the faces which follow being known, and consequently the opposite ones that precede, we should likewise know the determination of their motions. And thus we might find both the quantity and the determination of this circular motion, even in an immense vacuum, where there was nothing external or sensible with which the globes could be compared. But now, if in that space some remote bodies were placed that kept always a given position one to another, as the fixed stars do in our regions, we could not indeed determine from the relative translation of the globes among those bodies, whether the motion did belong to the globes or to the bodies. But if we observed the cord, and found that its tension was that very tension which the motions of the globes required, we might conclude the motion to be in the globes, and the bodies to be at rest; and then, lastly, from the translation of the globes among the bodies, we should find the determination of their motions. But how we are to obtain the true motions from their causes, effects, and apparent differences, and the converse, shall be explained more at large in the following treatise. For to this end it was that I composed it.

2 FROM *Ernst Mach* *The Science of Mechanics*

The growing list of discoveries and the steady advance of theory in the nineteenth century served as evidence to almost everyone that physics was undoubtedly on the right track. Only a very few sensed the growing weaknesses in the foundations of physics. Of these, perhaps the most acute was the Austrian physicist and philosopher Ernst Mach (1838–1916). In his The Science of Mechanics, A Critical and Historical Account of its Development, *first published in German in 1883, Mach challenged the Newtonian view of absolute space and time and insisted upon their relativity. Mach's influence was later acknowledged by Albert Einstein.*

NEWTON'S VIEWS OF TIME, SPACE AND MOTION

1. In a scholium which he appends immediately to his definitions, Newton presents his views regarding time and space which we must examine more in detail. We shall literally cite, to this end, only the passages that are absolutely necessary to the characterization of Newton's views*. . . .

2. It would appear as though Newton in the remarks here cited still stood under the influence of the mediæval philosophy,

*Mach here cites verbatim the beginning of the previous selection defining absolute and relative time. See pp. 9–11.

SOURCE. Ernst Mach, *The Science of Mechanics*, La Salle, Ill.: The Open Court Publishing Co., 1942, pp. 271–275, 276, 279–286, and 288–290.

as though he had grown unfaithful to his resolves to investigate only actual facts. When we say a thing *A* changes with the time, we mean simply that the conditions that determine a thing *A* depend on the conditions that determine another thing *B*. The vibrations of a pendulum take place *in time* when its excursion *depends* on the position of the earth. Since, however, in the observation of the pendulum, we are not under the necessity of taking into account its dependence on the position of the earth, but may compare it with any other thing (the conditions of which of course also depend on the position of the earth), the illusory notion easily arises that *all* the things with which we compare it are unessential. Nay, we may, in attending to the motion of a pendulum, neglect entirely other external things, and find that for every position of it our thoughts and sensations are different. Time, accordingly, appears to be some particular and independent thing, on the progress of which the position of the pendulum depends, while the things that we resort to for comparison and choose at random appear to play a wholly collateral part. But we must not forget that all things in the world are connected with one another and depend on one another, and that we ourselves and all our thoughts are also a part of nature. It is utterly beyond our power to *measure* the changes of things by *time*. Quite the contrary, time is an abstraction, at which we arrive by means of the changes of things; made because we are not restricted to any one *definite* measure, all being interconnected. A motion is termed uniform in which equal increments of space described correspond to equal increments of space described by some motion with which we form a comparison, as the rotation of the earth. A motion may, with respect to another motion, be uniform. But the question whether a motion is *in itself* uniform, is senseless. With just as little justice, also, may we speak of an "absolute time"—*of a time independent of* change. This absolute time can be measured by comparison with no motion; it has therefore neither a practical nor a scientific value; and no one is justified in saying that he knows aught about it. It is an idle metaphysical conception. . . .

We arrive at the idea of time—to express it briefly and popularly—by the connection of that which is contained in the province of our memory with that which is contained in the province of our sense-perception. When we say that time flows on in a definite direction or sense, we mean that physical events generally (and therefore also physiological events) take place only in a definite sense. Differences of temperature, electrical differences, differences of level generally, if left to themselves, all grow less and not greater. If we contemplate two bodies of different temperatures, put in contact and left wholly to themselves, we shall find that it is possible only for greater differences of temperature in the field of memory to exist with lesser ones in the field of sense-perception, and not the reverse. In all this there is simply expressed a peculiar and profound connection of things. To demand at the present time a full elucidation of this matter, is to anticipate, in the manner of speculative philosophy, the results of all future special investigation, that is, a perfected physical science. . . .

3. Views similar to those concerning time, are developed by Newton with respect to space and motion. . . .

If, in a material spatial system, there are masses with different velocities, which can enter into mutual relations with one another, these masses present to us forces. We can only decide how great these forces are when we know the velocities to which those masses are to be brought, *Resting* masses too are forces if *all* the masses do not rest. Think, for example, of Newton's rotating bucket in which the water is not yet rotating. If the mass m has the velocity v_1 and it is to be brought to the velocity v_2, the force which is to be spent on it is $p = m\ (v_1 - v_2)/t$, or the work which is to be expended is $ps = m(v_1^2 - v_2^2)$. *All* masses and *all* velocities, and consequently *all* forces, are relative. There is no decision about relative and absolute which we can possibly meet, to which we are forced, or from which we can obtain any intellectual or other advantage. When quite modern authors let themselves be led astray by the Newtonian arguments which are derived from the bucket of water, to distinguish between relative and absolute motion, they do not reflect that the system of the world is only given *once* to us, and the Ptolemaic or Copernican view is *our* interpretation, but both are equally actual. Try to fix Newton's bucket and rotate the heaven of fixed stars and then prove the absence of centrifugal forces.

4. It is scarcely necessary to remark that in the reflections here

presented Newton has again acted contrary to his expressed intention only to investigate *actual facts.* No one is competent to predicate things about absolute space and absolute motion; they are pure things of thought, pure mental constructs, that cannot be produced in experience. All our principles of mechanics are, as we have shown in detail, experimental knowledge concerning the relative positions and motions of bodies. Even in the provinces in which they are now recognized as valid, they could not, and were not, admitted without previously being subjected to experimental tests. No one is warranted in extending these principles beyond the boundaries of experience. In fact, such an extension is meaningless, as no one possesses the requisite knowledge to make use of it.

We must suppose that the change in the point of view from which the system of the world is regarded which was initiated by Copernicus, left deep traces in the thought of Galileo and Newton. But while Galileo, in his theory of the tides, quite naïvely chose the sphere of the fixed stars as the basis of a new system of coördinates, we see doubts expressed by Newton as to whether a given fixed star is at rest only apparently or really (*Principia*, 1687, p. 11). This appeared to him to cause the difficulty of distinguishing between true (absolute) and apparent (relative) motion. By this he was also impelled to set up the conception of *absolute space*. By further investigations in this direction—the discussion of the experiment of the rotating spheres which are connected together by a cord and that of the rotating water-bucket (pp. 9, 11)—he believed that he could prove an absolute rotation, though he could not prove any absolute translation. By absolute rotation he understood a rotation relative to the fixed stars, and here centrifugal forces can always be found. "But how we are to collect," says Newton in the Scholium at the end of the Definitions, "the true motions from their causes, effects, and apparent differences, and *vice versa*; how from the motions, either true or apparent, we may come to the knowledge of their causes and effects, shall be explained more at large in the following Tract." The resting sphere of fixed stars seems to have made a certain impression on Newton as well. The natural system of reference is for him that which has any uniform motion or translation without rotation (relatively to the sphere of fixed stars).

But do not the words quoted in inverted commas give the impression that Newton was glad to be able now to pass over to less precarious questions that could be tested by experience?

Let us look at the matter in detail. When we say that a body K alters its direction and velocity solely through the influence of another body K', we have asserted a conception that it is impossible to come at unless other bodies A, B, C . . . are present with reference to which the motion of the body K has been estimated. In reality, therefore, we are simply cognizant of a relation of the body K to A, B, C. . . . If now we suddenly neglect A, B, C . . . and attempt to speak of the deportment of the body K in absolute space, we implicate ourselves in a twofold error. In the first place, we cannot know how K would act in the absence of A, B, C . . . ; and in the second place, every means would be wanting of forming a judgment of the behavior of K and of putting to the test what we had predicted–which latter therefore would be bereft of all scientific significance. . . .

The motion of a body K can only be estimated by reference to other bodies A, B, C . . . . But since we always have at our disposal a sufficient number of bodies, that are as respects each other relatively fixed, or only slowly change their positions, we are, in such reference, restricted to no one *definite* body and can alternately leave out of account now this one and now that one. In this way the conviction arose that these bodies are indifferent generally.

It might be, indeed, that the isolated bodies A, B, C . . . play merely a collateral rôle in the determination of the motion of the body K, and that this motion is determined by a *medium* in which K exists. In such a case we should have to substitute this medium for Newton's absolute space. Newton certainly did not entertain this idea. Moreover, it is easily demonstrable that the atmosphere is not this motion-determinative medium. We should, therefore, have to picture to ourselves some other medium, filling, say, all space, with respect to the constitution of which and its kinetic relations to the bodies placed in it we have at present no adequate knowledge. In itself such a state of things would not belong to the impossibilities. It is known, from recent hydrodynamical investigations, that a rigid body experiences resistance in a frictionless fluid only when its velocity *changes*. True, this result is derived theoretically from the notion of inertia; but it might, conversely, also be regarded as the primitive fact from which we have to start. Although, practically, and at present, nothing is to be accomplished with this conception, we might still hope to learn more in the future concerning this hypothetical medium; and from the point of view of science it would be in every respect a more valuable acquisition than the forlorn idea of absolute space. When we reflect that we cannot abolish the isolated bodies $A, B, C \ldots$, that is, cannot determine by experiment whether the part they play is fundamental or collateral, that hitherto they have been the sole and only competent means of the orientation of motions and of the description of mechanical facts, it will be found expedient provisionally to regard all motions as determined by these bodies.

5. Let us now examine the point on which Newton, apparently with sound reasons, rests his distinction of absolute and relative motion. If the earth is affected with an *absolute* rotation about its axis, centrifugal forces are set up in the earth: it assumes an oblate form, the acceleration of gravity is diminished at the equator, the plane of Foucault's pendulum rotates, and so on. All these phenomena disappear if the earth is at rest and the other heavenly bodies are affected with absolute motion round it, such that the same *relative* rotation is produced. This is, indeed, the case, if we start *ab initio* from the idea of absolute space. But if we take our stand on the basis of facts, we shall find we have knowledge only of *relative* spaces and motions. *Relatively*, not considering the unknown and neglected medium of space, the motions of the universe are the same whether we adopt the Ptolemaic or the Copernican mode of view. Both views are, indeed, equally *correct;* only the latter is more simple and more *practical.* The universe is not *twice* given, with an earth at rest and an earth in motion; but only *once*, with its *relative* motions, alone determinable. It is, accordingly, not permitted us to say how things would be if the earth did not rotate. We may interpret the one case that is given us, in different ways. If, however, we so interpret it that we come into conflict with experience, our interpretation is simply wrong. The principles of mechanics can, indeed, be so conceived, that even for relative rotations centrifugal forces arise. ...

Newton's experiment with the rotating vessel of water simply informs us, that the relative rotation of the water with respect to the sides of the vessel produces *no* noticeable centrifugal forces, but that such forces *are* produced by its relative rotation with respect to the mass of the earth and the other celestial bodies. No one is competent to say how the experiment would turn out if the sides of the vessel increased in thickness and mass till they were ultimately several leagues thick. The one experiment only lies before us, and our business is, to bring it into accord with the other facts known to us, and not with the arbitrary fictions of our imagination. . . .

The comportment of terrestrial bodies with respect to the earth is reducible to the comportment of the earth with respect to the remote heavenly bodies. If we were to assert that we knew more of moving objects than this their last-mentioned, experimentally-given comportment with respect to the celestial bodies, we should render ourselves culpable of a falsity. When, accordingly, we say, that a body preserves unchanged its direction and velocity *in space*, our assertion is nothing more or less than an abbreviated reference to *the entire universe*. The use of such an abbreviated expression is permitted the original author of the principle, because he knows, that as things are no difficulties stand in the way of carrying out its implied directions. But no remedy lies in his power, if difficulties of the kind mentioned present themselves; if, for example, the requisite, relatively fixed bodies are wanting. . . .

8. The considerations just presented show, that it is not necessary to refer the law of inertia to a special absolute space. On the contrary, it is perceived that the masses that in the common phraseology exert forces on each other as well as those that exert none, stand with respect to acceleration in quite similar relations. We may, indeed, regard *all* masses as related to each other. That *accelerations* play a prominent part in the relations of the masses, must be accepted as a fact of experience; which does not, however, exclude attempts to *elucidate* this fact by a comparison of it with other facts, involving the discovery of new points of view. In all the processes of nature the *differences* of certain quantities u play a determinative rôle. Differences of temperature, of potential function, and so forth, induce the natural processes, which consist

in the equalization of these differences. The familiar expressions d^2u/dx^2, d^2u/dy^2, d^2u/dz^2, which are determinative of the character of the equalization, may be regarded as the measure of the departure of the condition of any point from the mean of the conditions of its environment—to which mean the point tends. The accelerations of masses may be analogously conceived. The great distances between masses that stand in no especial force-relation to one another, change *proportionately to each other.* If we lay off, therefore, a certain distance ρ as abscissa, and another r as ordinate, we obtain a straight line (Fig. 143). Every r-ordinate corresponding to a definite ρ-value represents, accord-

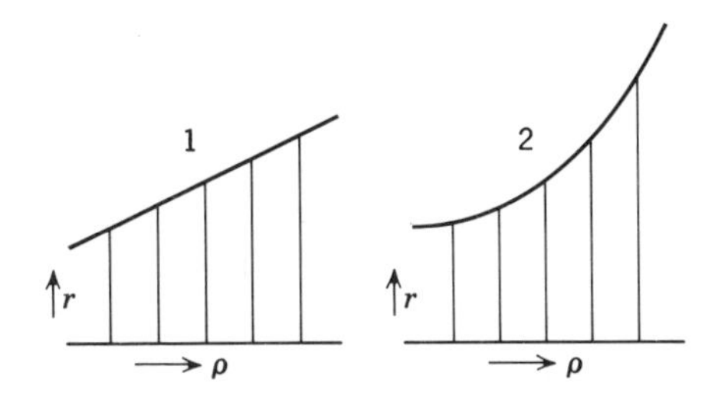

Fig. 143

ingly, the mean of the adjacent ordinates. If a force-relation exists between the bodies, some value d^2r/dt^2 is determined by it which conformably to the remarks above we may replace by an expression of the form $d^2r/d\rho^2$. By the force-relation, therefore, a *departure* of the r-ordinate from the *mean of the adjacent ordinates* is produced, which would not exist if the supposed force-relation did not obtain. This intimation will suffice here.

9. We have attempted in the foregoing to give the law of inertia a different expression from that in ordinary use. This expression will, so long as a sufficient number of bodies are apparently fixed in space, accomplish the same as the ordinary one. It is as easily applied, and it encounters the same difficulties. In the one case we are unable to come at an absolute space, in the other a limited number of masses only is within the reach of our knowl-

edge, and the summation indicated can consequently not be fully carried out. It is impossible to say whether the new expression would still represent the true condition of things if the stars were to perform rapid movements among one another, The general experience cannot be constructed from the particular case given us. We must, on the contrary, *wait* until such an experience presents itself. Perhaps when our physico-astronomical knowledge has been extended, it will be offered somewhere in celestial space, where more violent and complicated motions take place than in our environment. The most important result of our reflections is, however, *that precisely the apparently simplest mechanical principles are of a very complicated character, that these principles are founded on uncompleted experiences, nay on experiences that never can be fully completed, that practically, indeed, they are sufficiently secured, in view of the tolerable stability of our environment, to serve as the foundation of mathematical deduction, but that they can by no means themselves be regarded as mathematically established truths but only as principles that not only admit of constant control by experience but actually require it.*

3 *Albert A. Michelson and Edward W. Morley* *"On the Relative Motion of the Earth and the Luminiferous Ether"*

In 1887 Albert A. Michelson and Edward W. Morley performed an experiment in Cleveland, Ohio, which has since become a classic. The Michelson-Morley experiment was designed to detect the "ether wind" that physical theory predicted should exist. The results were negative. As the authors point out in their concluding statement, such results seriously undermined current theories of the relation of the motion of the earth to that of the ether.

SOURCE. Albert A. Michelson and Edward W. Morley, "On the Relative Motion of the Earth and the Luminiferous Ether," *American Journal of Science*, Third Series, Vol. 34, 1887, pp. 333–341. Reprinted by permission of the publisher.

The Michelson-Morley experiment is more often cited than read today. Did it really *disprove the existence of an "ether wind" or were the number of "runs" too small to determine this conclusively? Could experimental errors account for the failure to detect the "wind"? In short, did the Michelson-Morley experiment really force a rethinking of the foundations of physics?*

The discovery of the aberration of light was soon followed by an explanation according to the emission theory. The effect was attributed to a simple composition of the velocity of light with the velocity of the earth in its orbit. The difficulties in this apparently sufficient explanation were overlooked until after an explanation on the undulatory theory of light was proposed. This new explanation was at first almost as simple as the former. But it failed to account for the fact proved by experiment that the aberration was unchanged when observations were made with a telescope filled with water. For if the tangent of the angle of aberration is the ratio of the velocity of the earth to the velocity of light, then, since the latter velocity in water is three-fourths its velocity in a vacuum, the aberration observed with a water telescope should be four-thirds of its true value.

On the undulatory theory, according to Fresnel, first, the ether is supposed to be at rest except in the interior of transparent media, in which secondly, it is supposed to move with a velocity less than the velocity of the medium in the ratio $\frac{(n^2-1)}{n^2}$, where n is the index of refraction. These two hypotheses give a complete and satisfactory explanation of aberration. The second hypothesis, notwithstanding its seeming improbability, must be considered as fully proved, first, by the celebrated experiment of Fizeau, and secondly, by the ample confirmation of our own work. The experimental trial of the first hypothesis forms the subject of the present paper.

If the earth were a transparent body, it might perhaps be conceded, in view of the experiments just cited, that the intermolecular ether was at rest in space, notwithstanding the motion of the earth in its orbit; but we have no right to extend the conclusion from these experiments to opaque bodies. But there can hardly be question that the ether can and does pass through metals. Lorentz cites the illustration of a metallic barometer tube. When the tube is inclined the ether in the space above the mercury is certainly forced out, for it is incompressible. But again we have no right to assume that it makes its escape with perfect freedom, and if there be any resistance, however slight, we certainly could not assume an opaque body such as the whole earth to offer free passage through its entire mass. . . .

In April, 1881, a method was proposed and carried out for testing the question experimentally.

In deducing the formula for the quantity to be measured, the effect of the motion of the earth through the ether on the path of the ray at right angles to this motion was overlooked. The discussion of this oversight and of the entire experiment forms the subject of a very searching analysis by H. A. Lorentz, who finds that this effect can by no means be disregarded. In consequence, the quantity to be measured had in fact but one-half the value supposed, and as it was already barely beyond the limits of errors of experiment, the conclusion drawn from the result of the experiment might well be questioned; since, however, the main portion of the theory remains unquestioned, it was decided to repeat the experiment with such modifications as would insure a theoretical result much too large to be masked by experimental errors. The theory of the method may be briefly stated as follows:

Let *sa*, Fig. 1, be a ray of light which is partly reflected in *ab*, and partly transmitted in *ac*, being returned by the mirrors *b* and *c*, along *ba* and *ca*. *ba* is partly transmitted along *ad*, and *ca* is partly reflected along *ad*. If then the paths *ab* and *ac* are equal, the two rays interfere along *ad*. Suppose now, the ether being at rest, that the whole apparatus moves in the direction *sc*, with the velocity of the earth in its orbit, the directions and distances traversed by the rays will be altered thus: The ray *sa* is reflected along *ab*, Fig. 2; the angle *bab*, being equal to the aberration $= a$, is returned along *ba′*, ($aba' = 2a$), and goes to the focus of the telescope, whose direction is unaltered. The transmitted ray goes along *ac*, is returned along *ca′*, and is reflected at *a′*, making *ca′e* equal 90 − *a*, and therefore still coinciding with the first ray. It may be remarked that the rays *ba′* and *ca′*, do not now meet exactly in the same point *a′*, though the difference is of the second order; this does not affect the validity of the reasoning. Let it now be required to find the difference in the two paths *aba′*, and *aca′*. Let V = velocity of light.

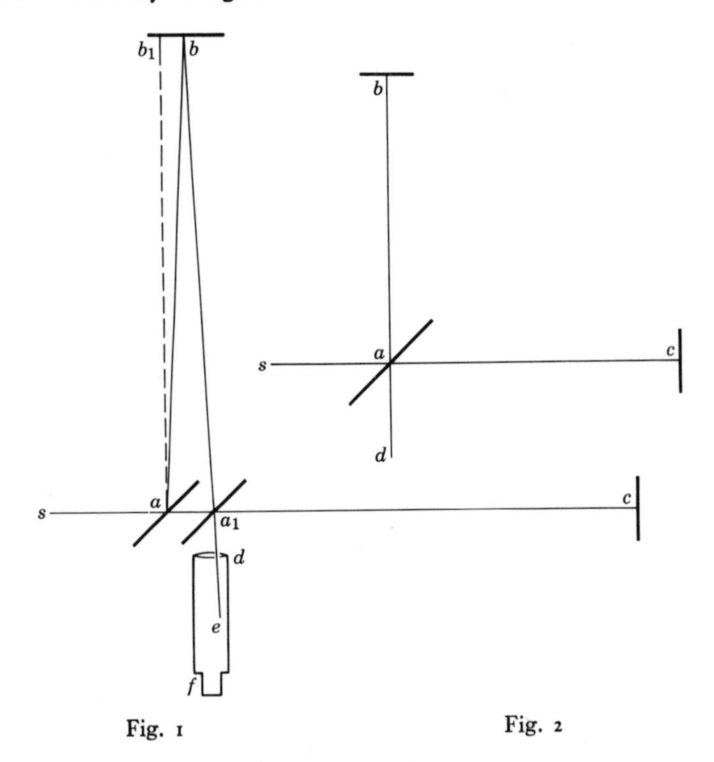

Fig. 1 Fig. 2

v = velocity of the earth in its orbit.
D = distance *ab* or *ac* (Fig. 1).
T = time light occupies to pass from *a* to *c*.
T_1 = time light occupies to return from *c* to *a′* (Fig. 2).

Then

$$T = \frac{D}{V-v}, \qquad T_1 = \frac{D}{V+v}.$$

The whole time of going and coming is

$$T + T_1 = 2D\frac{V}{V^2 - v^2},$$

and the distance traveled in this time is

$$2D\frac{V^2}{V^2 - v^2} = 2D\left(1 + \frac{v^2}{V^2}\right),$$

neglecting terms of the fourth order. The length of the other path is evidently $2D\sqrt{1 + v^2/V^2}$, or to the same degree of accuracy, $2D(1 + v^2/2V^2)$. The difference is therefore $D(v^2/V^2)$; If now the whole apparatus be turned through 90°, the difference will be in the opposite direction, hence the displacement of the interference fringes should be $2D(v^2/V^2)$. Considering only the velocity of the earth in its orbit, this would be $2D \times 10^{-8}$. If, as was the case in the first experiment, $D = 2 \times 10^6$ waves of yellow light, the displacement to be expected would be 0.04 of the distance between the interference fringes.

In the first experiment one of the principal difficulties encountered was that of revolving the apparatus without producing distortion; and another was its extreme sensitiveness to vibration. This was so great that it was impossible to see the interference fringes except at brief intervals when working in the city, even at two o'clock in the morning. Finally, as before remarked, the quantity to be observed, namely, a displacement of something less than a twentieth of the distance between the interference fringes, may have been too small to be detected when masked by experimental errors.

The first named difficulties were entirely overcome by mounting the apparatus on a massive stone floating on mercury; and the second by increasing, by repeated reflection, the path of the light to about ten times its former value.

The apparatus is represented in perspective in Fig. 3, in plan in Fig. 4, and in vertical section in Fig. 5. The stone *a* (Fig. 5) is about 1.5 meter square and 0.3 meter thick. It rests on an annular wooden float *bb*, 1.5 meter outside diameter, 0.7 meter inside diameter, and 0.25 meter thick. The float rests on mercury contained in the cast-iron trough *cc*, 1.5 centimeter thick, and of such dimensions as to leave a clearance of about one centimeter around the float. A pin *d*, guided by arms *gggg*, fits into a socket *e* attached to the float. The pin may be pushed into the socket or be withdrawn, by a lever pivoted at *f*. This pin keeps the float concentric with the trough, but does not bear any part of the weight of the stone. The annular iron trough rests on a bed of cement on a low brick pier built in the form of a hollow octagon.

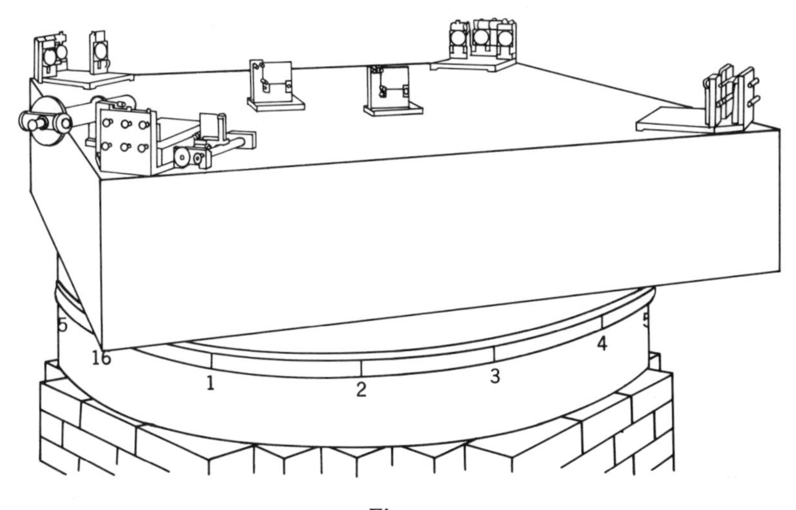

Fig. 3

At each corner of the stone were placed four mirrors dd_1 ee_1 (Fig. 4). Near the center of the stone was a plane-parallel glass *b*. These were so disposed that light from an argand burner *a*, passing through a lens, fell on *b* so as to be in part reflected to d_1, the two pencils followed the paths indicated in the figure, *bdedbf* and $bd_1e_1d_1bf$ respectively, and were observed by the telescope *f*. Both *f* and *a* revolved with the stone. The mirrors were of speculum metal carefully worked to optically plane surfaces five centimeters in diameter, and the glasses *b* and *c* were plane-parallel and of the same thickness, 1.25 centimeter; their surfaces measured 5.0 by 7.5 centimeters. The second of these was placed in the path of one of the pencils to compensate for the passage of the other through the same thickness of glass. The whole of the optical portion of the apparatus was kept covered with a wooden cover to prevent air currents and rapid changes of temperature.

The adjustment was effected as follows: The mirrors having been adjusted by screws in the castings which held the mirrors, against which they were pressed by springs, till light from both

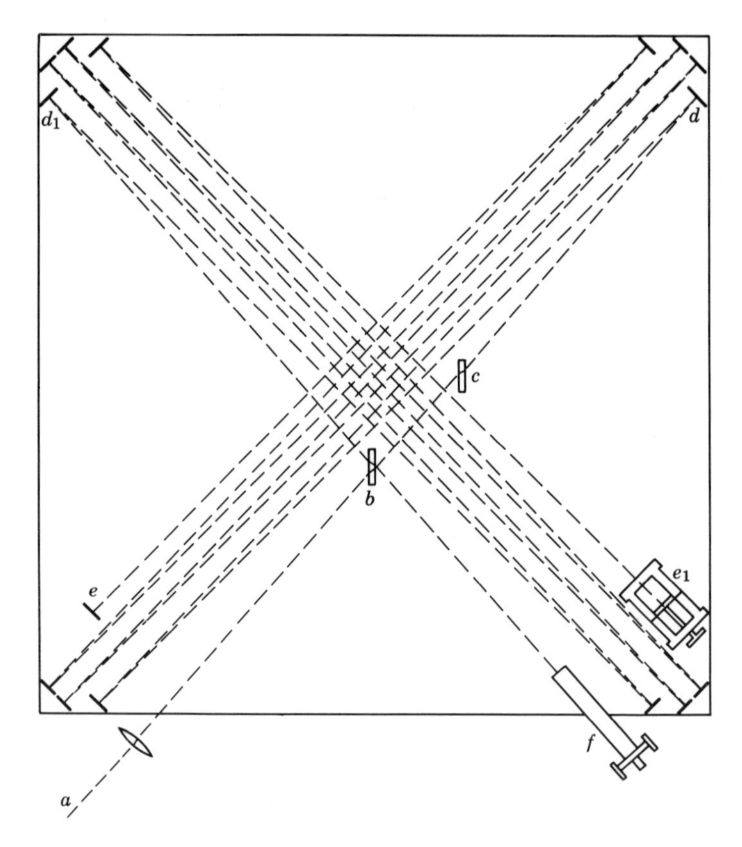

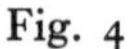
Fig. 4

pencils could be seen in the telescope, the lengths of the two paths were measured by a light wooden rod reaching diagonally from mirror to mirror, the distance being read from a small steel scale to tenths of millimeters. The difference in the lengths of the two paths was then annulled by moving the mirror e_1. This mirror had three adjustments; it had an adjustment in altitude and one in azimuth, like all the other mirrors, but finer; it also had an adjustment in the direction of the incident ray, sliding forward or backward, but keeping very accurately parallel to its former plane. The three adjustments of this mirror could be made with the wooden cover in position.

The paths being now approximately equal, the two images of the source of light or of some well-defined object placed in front of the condensing lens, were made to coincide, the telescope was now adjusted for distinct vision of the expected interference bands, and sodium light was substituted for white light, when the interference bands appeared. These were now made as clear as possible by adjusting the mirror $e_{,}$; then white light was restored, the screw altering the length of path was very slowly

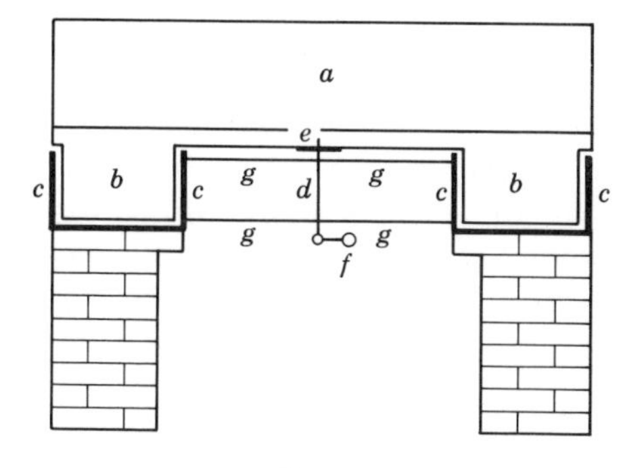

Fig. 5

moved (one turn of a screw of one hundred threads to the inch altering the path nearly 1000 wave-lengths) till the colored interference fringes reappeared in white light. These were now given a convenient width and position, and the apparatus was ready for observation.

The observations were conducted as follows: Around the cast-iron trough were sixteen equidistant marks. The apparatus was revolved very slowly (one turn in six minutes) and after a few minutes the cross wire of the micrometer was set on the clearest of the interference fringes at the instant of passing one of the marks. The motion was so slow that this could be done readily and accurately. The reading of the screw-head on the micrometer was noted, and a very slight and gradual impulse was given to keep up the motion of the stone; on passing the second mark, the same process was repeated, and this was continued till the apparatus had completed six revolutions. It was found that by keeping the apparatus in slow uniform motion, the results were much more uniform and consistent than when the stone was brought to rest for every observation; for the effects of strains could be noted for at least half a minute after the stone came to rest, and during this time effects of change of temperature came into action.

The following tables give the means of the six readings; the first, for observations made near noon, the second, those near six o'clock in the evening. The readings are divisions of the screw-heads. The width of the fringes varied from 40 to 60 divisions, the mean value being near 50, so that one division means 0.02 wave-length. The rotation in the observations at noon was contrary to, and in the evening observations, with, that of the hands of a watch.

The results of the observations are expressed graphically in Fig. 6. The upper is the curve for the observations at noon, and the lower that for the evening observations. The dotted curves represent *one-eighth* of the theoretical displacements. It seems fair to conclude from the figure that if there is any displacement

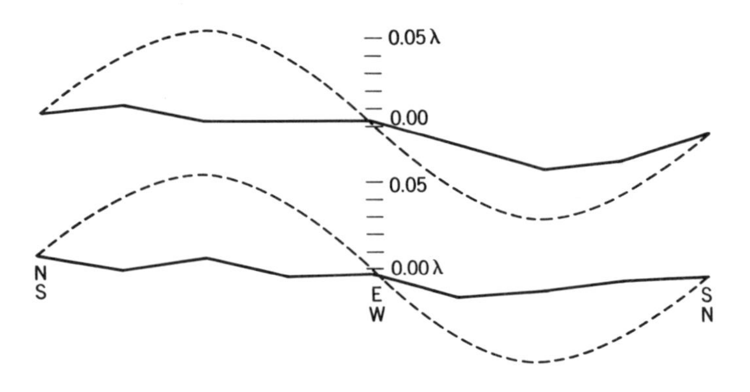

Fig. 6

due to the relative motion of the earth and the luminiferous ether, this cannot be much greater than 0.01 of the distance between the fringes.

Noon Observations

	16.	1.	2.	3.	4.	5.	6.	7.	8.	9.	10.	11.	12.	13.	14.	15.	16.
July 8	44.7	44.0	43.5	39.7	35.2	34.7	34.3	32.5	28.2	26.2	23.8	23.2	20.3	18.7	17.5	16.8	13.7
July 9	57.4	57.3	58.2	59.2	58.7	60.2	60.8	62.0	61.5	63.3	65.8	67.3	69.7	70.7	73.0	70.2	72.2
July 11	27.3	23.5	22.0	19.3	19.2	19.3	18.7	18.8	16.2	14.3	13.3	12.8	13.3	12.3	10.2	7.3	6.5
Mean	43.1	41.6	41.2	39.4	37.7	38.1	37.9	37.8	35.3	34.6	34.3	34.4	34.4	33.9	33.6	31.4	30.8
Mean in w. l.	.862	.832	.824	.788	.754	.762	.758	.756	.706	.692	.686	.688	.688	.678	.672	.628	.616
	.706	.692	.686	.688	.688	.678	.672	.628	.616								
Final mean	.784	.762	.755	.738	.721	.720	.715	.692	.661								

P. M. Observations

	16.	1.	2.	3.	4.	5.	6.	7.	8.	9.	10.	11.	12.	13.	14.	15.	16.
July 8	61.2	63.3	63.3	68.2	67.7	69.3	70.3	69.8	69.0	71.3	71.3	70.5	71.2	71.2	70.5	72.5	75.7
July 9	26.0	36.0	28.2	29.2	31.5	32.0	31.3	31.7	33.0	35.8	36.5	37.3	38.8	41.0	42.7	43.7	44.0
July 12	66.8	66.5	66.0	64.3	62.2	61.0	61.3	59.7	58.2	55.7	53.7	54.7	55.0	58.2	58.5	57.0	56.0
Mean	51.3	51.9	52.5	53.9	53.8	54.1	54.3	53.7	53.4	54.3	53.8	54.2	55.0	56.8	57.2	57.7	58.6
Mean in w. l.	1.026	1.038	1.050	1.078	1.076	1.082	1.086	1.074	1.068	1.086	1.076	1.084	1.100	1.136	1.144	1.154	1.172
	1.068	1.086	1.076	1.084	1.100	1.136	1.144	1.154	1.172								
Final mean	1.047	1.062	1.063	1.081	1.088	1.109	1.115	1.114	1.120								

Considering the motion of the earth in its orbit only, this displacement should be

$$2D\frac{v^2}{V^2} = 2D \times 10^{-8}.$$

The distance D was about eleven meters, or 2×10^7 wave-lengths of yellow light; hence the displacement to be expected was 0.4 fringe. The actual displacement was certainly less than the twentieth part of this, and probably less than the fortieth part. But since the displacement is proportional to the square of the velocity, the relative velocity of the earth and the ether is probably less than one-sixth the earth's orbital velocity, and certainly less than one-fourth.

In what precedes, only the orbital motion of the earth is considered. If this is combined with the motion of the solar system, concerning which but little is known with certainty, the result would have to be modified; and it is just possible that the resultant velocity at the time of the observations was small though the chances are much against it. The experiment will therefore be repeated at intervals of three months, and thus all uncertainty will be avoided. [It was not repeated. Ed.]

It appears, from all that precedes, reasonably certain that if there be any relative motion between the earth and the luminiferous ether, it must be small; quite small enough entirely to refute Fresnel's explanation of aberration. Stokes has given a theory of aberration which assumes the ether at the earth's surface to be at rest with regard to the latter, and only requires in addition that the relative velocity have a potential; but Lorentz shows that these conditions are incompatible. Lorentz then proposes a modification which combines some ideas of Stokes and Fresnel, and assumes the existence of a potential, together with Fresnel's coefficient. If now it were legitimate to conclude from the present work that the ether is at rest with regard to the earth's surface, according to Lorentz there could not be a velocity potential, and his own theory also fails.

4 H. A. Lorentz
"Michelson's Interference Experiment"

One of the first to come to grips with the implications of the Michelson-Morley experiment was the great Dutch theoretical physicist H. A. Lorentz (1853–1928). He was able to save the ether by postulating a change in the dimensions of measuring rods as they moved through the ether. This was the origin of the Lorentz-Fitzgerald contraction.

Lorentz's paper on Michelson's experiment was published in Leyden, Holland, in 1895, ten years before Einstein created the Special Theory of Relativity. Einstein later remarked that he had not heard of Lorentz's paper when he wrote his 1905 article on the motion of electrodynamic bodies in which special relativity was introduced.

---

1. As Maxwell first remarked and as follows from a very simple calculation, the time required by a ray of light to travel from a point A to a point B and back to A must vary when the two points together undergo a displacement without carrying the ether with them. The difference is, certainly, a magnitude of second order; but it is sufficiently great to be detected by a sensitive interference method.

The experiment was carried out by Michelson in 1881. His apparatus, a kind of interferometer, had two horizontal arms, P and Q, of equal length and at right angles one to the other. Of the two mutually interfering rays of light the one passed along the arm P and back, the other along the arm Q and back. The whole instrument, including the source of light and the arrangement for taking observations, could be revolved about a vertical axis; and those two positions come especially under consideration in which the arm P or the arm Q lay as nearly as possible in the direction of the Earth's motion. On the basis of Fresnel's theory

SOURCE. H. A. Lorentz, A. Einstein, H. Minkowski, and H. Weyl, *The Principle of Relativity*, W. Perrett and G. B. Jeffery, trs., New York: Dover Publications, Inc., 1923, pp. 3–7. Reprinted with the permission of the publisher.

it was anticipated that when the apparatus was revolved from one of these *principle positions* into the other there would be a displacement of the interference fringes.

But of such a displacement—for the sake of brevity we will call it the Maxwell displacement—conditioned by the change in the times of propagation, no trace was discovered, and accordingly Michelson thought himself justified in concluding that while the Earth is moving, the ether does not remain at rest. The correctness of this inference was soon brought into question, for by an oversight Michelson had taken the change in the phase difference, which was to be expected in accordance with the theory, at twice its proper value. If we make the necessary correction, we arrive at displacements no greater than might be masked by errors of observation.

Subsequently Michelson took up the investigation anew in collaboration with Morley, enhancing the delicacy of the experiment by causing each pencil to be reflected to and fro between a number of mirrors, thereby obtaining the same advantage as if the arms of the earlier apparatus had been considerably lengthened. The mirrors were mounted on a massive stone disc, floating on mercury, and therefore easily revolved. Each pencil now had to travel a total distance of 22 meters, and on Fresnel's theory the displacement to be expected in passing from the one principal position to the other would be 0.4 of the distance between the interference fringes. Nevertheless the rotation produced displacements not exceeding 0.02 of this distance, and these might well be ascribed to errors of observation.

Now, does this result entitle us to assume that the ether takes part in the motion of the Earth, and therefore that the theory of aberration given by Stokes is the correct one? The difficulties which this theory encounters in explaining aberration seem too great for me to share this opinion, and I would rather try to remove the contradiction between Fresnel's theory and Michelson's result. An hypothesis which I brought forward some time ago, and which, as I subsequently learned, has also occurred to Fitzgerald, enables us to do this. The next paragraph will set out this hypothesis.

2. To simplify matters we will assume that we are working with apparatus as employed in the first experiments, and that in

the one principal position the arm P lies exactly in the direction of the motion of the Earth. Let v be the velocity of this motion, L the length of either arm, and hence $2L$ the path traversed by the rays of light. According to the theory, the turning of the apparatus through 90° causes the time in which the one pencil travels along P and back to be longer than the time which the other pencil takes to complete its journey by

$$\frac{Lv^2}{c^3}$$

There would be this same difference if the translation had no influence and the arm P were longer than the arm Q by $\frac{1}{2}Lv^2/c^2$. Similarly with the second principal position.

Thus we see that the phase differences expected by the theory might also arise if, when the apparatus is revolved, first the one arm and then the other arm were the longer. It follows that the phase differences can be compensated by contrary changes of the dimensions.

If we assume the arm which lies in the direction of the Earth's motion to be shorter than the other by $\frac{1}{2}Lv^2/c^2$, and, at the same time, that the translation has the influence which Fresnel's theory allows it, then the result of the Michelson experiment is explained completely.

Thus one would have to imagine that the motion of a solid body (such as a brass rod or the stone disc employed in the later experiments) through the resting ether exerts upon the dimensions of that body an influence which varies according to the orientation of the body with respect to the direction of motion. If, for example, the dimensions parallel to this direction were changed in the proportion of 1 to $1 + \delta$, and those perpendicular in the proportion of 1 to $1 + \epsilon$, then we should have the equation

$$\epsilon - \delta = \tfrac{1}{2}\frac{v^2}{c^2} \tag{1}$$

in which the value of one of the quantities δ and ϵ would remain undetermined. It might be that $\epsilon = 0$, $\delta = -\frac{1}{2}v^2/c^2$, but also $\epsilon = \frac{1}{2}v^2/c^2$, $\delta = 0$, or $\epsilon = \frac{1}{4}v^2/c^2$, and $\delta = -\frac{1}{4}v^2/c^2$.

3. Surprising as this hypothesis may appear at first sight, yet we shall have to admit that it is by no means far-fetched, as soon

as we assume that molecular forces are also transmitted through the ether, like the electric and magnetic forces of which we are able at the present time to make this assertion definitely. If they are so transmitted, the translation will very probably affect the action between two molecules or atoms in a manner resembling the attraction or repulsion between charged particles. Now, since the form and dimensions of a solid body are ultimately conditioned by the intensity of molecular actions, there cannot fail to be a change of dimensions as well.

From the theoretical side, therefore, there would be no objection to the hypothesis. As regards its experimental proof, we must first of all note that the lengthenings and shortenings in question are extraordinarily small. We have $v^2/c^2 = 10^{-8}$, and thus, if $\epsilon = 0$, the shortening of the one diameter of the Earth would amount to about 6.5 cm. The length of a meter rod would change, when moved from one principal position into the other, by about 1/200 micron. One could hardly hope for success in trying to perceive such small quantities except by means of an interference method. We should have to operate with two perpendicular rods, and with two mutually interfering pencils of light, allowing the one to travel to and fro along the first rod, and the other along the second rod. But in this way we should come back once more to the Michelson experiment, and revolving the apparatus we should perceive no displacement of the fringes. Reversing a previous remark, we might now say that the displacement produced by the alterations of length is compensated by the Maxwell displacement.

4. It is worth noticing that we are led to just the same changes of dimensions as have been presumed above if we, *firstly*, without taking molecular movement into consideration, assume that in a solid body left to itself the forces, attractions or repulsions, acting upon any molecule maintain one another in equilibrium, and, *secondly*—though to be sure, there is no reason for doing so—if we apply to these molecular forces the law which in another place we deduced for electrostatic actions. For if we now understand by S_1 and S_2 not, as formerly, two systems of charged particles, but two systems of molecules—the second at rest and the first moving with a velocity v in the direction of the axis of x—between the dimensions of which the relationship subsists as previously

stated; and if we assume that in both systems the x components of the forces are the same, while the y and z components differ from one another by the factor $\sqrt{1 - v^2/c^2}$, then it is clear that the forces in S_1 will be in equilibrium whenever they are so in S_2. If therefore S_2 is the state of equilibrium of a solid body at rest, then the molecules in S_1 have precisely those positions in which they can persist under the influence of translation. The displacement would naturally bring about this disposition of the molecules of its own accord, and thus effect a shortening in the direction of motion in the proportion of 1 to $\sqrt{1 - v^2/c^2}$, in accordance with the formulæ given in the above-mentioned paragraph. This leads to the values

$$\delta = -\tfrac{1}{2}\frac{v^2}{c^2}, \qquad \epsilon = 0$$

in agreement with (1).

In reality the molecules of a body are not at rest, but in every "state of equilibrium" there is a stationary movement. What influence this circumstance may have in the phenomenon which we have been considering is a question which we do not here touch upon; in any case the experiments of Michelson and Morley, in consequence of unavoidable errors of observation, afford considerable latitude for the values of δ and ϵ.

5 *Jules Henri Poincaré*
"The Principles of Mathematical Physics"

In 1904, only one year before Einstein's paper, at the occasion of the World's Fair celebrating the Louisiana Purchase one century before, the organizers of the St. Louis International Exposition sponsored an International Congress of the Arts and Sciences. The great

NB

SOURCE. Howard J. Rogers, ed., *Congress of Arts and Science*, Vol. I: Philosophy and Mathematics, George Bruce Halsted, tr., Boston: Houghton, Mifflin and Company, 1905, pp. 604, 608, 610–618, and 621–622. Reprinted by permission of Houghton Mifflin Company.

men in all fields were invited to come and describe the progress made in all areas of human thought during the nineteenth century. One of those invited was Henri Poincaré, the French mathematician and theoretical physicist; at fifty years of age, at the height of his influence. Poincaré, like Lorentz, sensed the unease in physics and tried to explain it to a lay audience.

What is the actual state of mathematical physics? What are the problems it is led to set itself? What is its future? Is its orientation on the point of modifying itself?

Will the aim and the methods of this science appear in ten years to our immediate successors in the same light as to ourselves; or, on the contrary, are we about to witness a profound transformation? Such are the questions we are forced to raise in entering to-day upon our investigation.

If it is easy to propound them, to answer is difficult. . . .

It is the mathematical physics of our fathers which has familiarized us little by little with these divers principles; which has taught us to recognize them under the different vestments in which they disguise themselves. One has to compare them to the data of experience, to find how it was necessary to modify their enunciation so as to adapt them to these data; and by these processes they have been enlarged and consolidated.

So we have been led to regard them as experimental verities; the conception of central forces became then a useless support, or rather an embarrassment, since it made the principles partake of its hypothetical character.

The frames have not therefore broken, because they were elastic; but they have enlarged; our fathers, who established them, did not work in vain, and we recognize in the science of to-day the general traits of the sketch which they traced.

Are we about to enter now upon the eve of a second crisis? Are these principles on which we have built all about to crumble away in their turn? . . .

We come to the principle of relativity: this not only is confirmed by daily experience, not only is it a necessary consequence of the hypothesis of central forces, but it is imposed in an irresistible way upon our good sense, and yet it also is battered.

Consider two electrified bodies; though they seem to us at rest, they are both carried along by the motion of the earth; an electric charge in motion, Rowland has taught us, is equivalent to a current; these two charged bodies are, therefore, equivalent to two parallel currents of the same sense and these two currents should attract each other. In measuring this attraction, we measure the velocity of the earth; not its velocity in relation to the sun or the fixed stars, but its absolute velocity.

I know it will be said that it is not its absolute velocity that is measured, but its velocity in relation to the ether. How unsatisfactory that is! Is it not evident that from a principle so understood we could no longer get anything? It could no longer tell us anything just because it would no longer fear any contradiction.

If we succeed in measuring anything, we should always be free to say that this is not the absolute velocity in relation to the ether, it might always be the velocity in relation to some new unknown fluid with which we might fill space.

Indeed, experience has taken on itself to ruin this interpretation of the principle of relativity; all attempts to measure the velocity of the earth in relation to the ether have led to negative results. This time experimental physics has been more faithful to the principle than mathematical physics; the theorists, to put in accord their other general views, would not have spared it; but experiment has been stubborn in confirming it.

The means have been varied in a thousand ways and finally Michelson has pushed precision to its last limits; nothing has come of it. It is precisely to explain this obstinacy that the mathematicians are forced to-day to employ all their ingenuity.

Their task was not easy, and if Lorentz has gotten through it, it is only by accumulating hypotheses.

The most ingenious idea has been that of local time.

Imagine two observers who wish to adjust their watches by optical signals; they exchange signals, but as they know that the transmission of light is not instantaneous, they take care to cross them.

When the station B perceives the signal from the station A, its clock should not mark the same hour as that of the station A at the moment of sending the signal, but this hour augmented by a constant representing the duration of the transmission. Suppose, for example, that the station A sends its signal when its clock marks the hour 0, and that the station B perceives it when its clock marks the hour t. The clocks are adjusted if the slowness equal to t represents the duration of the transmission, and to verify it the station B sends in its turn a signal when its clock marks 0; then the station A should perceive it when its clock marks t. The timepieces are then adjusted. And in fact, they mark the same hour at the same physical instant, but on one condition, namely, that the two stations are fixed. In the contrary case the duration of the transmission will not be the same in the two senses, since the station A, for example, moves forward to meet the optical perturbation emanating from B, while the station B flies away before the perturbation emanating from A. The watches adjusted in that manner do not mark, therefore, the true time; they mark what one may call the *local time*, so that one of them goes slow on the other. It matters little, since we have no means of perceiving it. All the phenomena which happen at A, for example, will be late, but all will be equally so, and the observer who ascertains them will not perceive it, since his watch is slow; so, as the principle of relativity would have it, he will have no means of knowing whether he is at rest or in absolute motion.

Unhappily, that does not suffice, and complementary hypotheses are necessary; it is necessary to admit that bodies in motion undergo a uniform contraction in the sense of the motion. One of the diameters of the earth, for example, is shrunk by 1/200,000,000 in consequence of the motion of our planet, while the other diameter retains its normal length. Thus, the last little differences find themselves compensated. And then there still is the hypothesis about forces. Forces, whatever be their origin, gravity as well as elasticity, would be reduced in a certain proportion in a world animated by a uniform translation; or, rather, this would happen for the components perpendicular to the translation; the components parallel would not change.

Resume, then, our example of two electrified bodies; these bodies repel each other, but at the same time if all is carried along in a uniform translation, they are equivalent to two parallel currents of the same sense which attract each other. This electrodynamic attraction diminishes, therefore, the electro-static repulsion, and the total repulsion is more feeble than if the two bodies were at rest. But since to measure this repulsion we must balance it by another force, and all these other forces are reduced in the same proportion, we perceive nothing.

Thus, all is arranged, but are all the doubts dissipated?

What would happen if one could communicate by non-luminous signals whose velocity of propagation differed from that of light? If, after having adjusted the watches by the optical procedure, one wished to verify the adjustment by the aid of these new signals, then would appear divergences which would render evident the common translation of the two stations. And are such signals inconceivable, if we admit with Laplace that universal gravitation is transmitted a million times more rapidly than light?

Thus, the principle of relativity has been valiantly defended in these latter times, but the very energy of the defense proves how serious was the attack.

Let us speak now of the principle of Newton, on the equality of action and reaction.

This is intimately bound up with the preceding, and it seems indeed that the fall of the one would involve that of the other. Thus we should not be astonished to find here the same difficulties.

Electrical phenomena, we think, are due to the displacements of little charged particles, called electrons, immersed in the medium that we call ether. The movements of these electrons produce perturbations in the neighboring ether; these perturbations propagate themselves in every direction with the velocity of light, and in turn other electrons, originally at rest, are made to vibrate when the perturbation reaches the parts of the ether which touch them.

The electrons, therefore, act upon one another, but this action is not direct, it is accomplished through the ether as intermediary.

Under these conditions can there be compensation between action and reaction, at least for an observer who should take account only of the movements of matter, that is to say, of the electrons, and who should be ignorant of those of the ether that he could not see? Evidently not. Even if the compensation should be exact, it could not be simultaneous. The perturbation is propagated with a finite velocity; it, therefore, reaches the second electron only when the first has long ago entered upon its rest.

This second electron, therefore, will undergo, after a delay, the action of the first, but certainly it will not react on this, since around this first electron nothing any longer budges.

The analysis of the facts permits us to be still more precise. Imagine for example, a Hertzian generator, like those employed in wireless telegraphy; it sends out energy in every direction; but we can provide it with a parabolic mirror, as Hertz did with his smallest generators, so as to send all the energy produced in a single direction.

What happens, then, according to the theory? It is that the appartus recoils as if it were a gun and as if the energy it has projected were a bullet; and that is contrary to the principle of Newton, since our projectile here has no mass, it is not matter, it is energy.

It is still the same, moreover, with a beacon light provided with a reflector, since light is nothing but a perturbation of the electromagnetic field. This beacon light should recoil as if the light it sends out were a projectile. What is the force that this recoil should produce? It is what one has called the Maxwell-Bartholdi pressure. It is very minute, and it has been difficult to put it into evidence even with the most sensitive radiometers; but it suffices that it exists.

If all the energy issuing from our generator falls on a receiver, this will act as if it had received a mechanical shock, which will represent in a sense the compensation of the recoil of the generator; the reaction will be equal to the action, but it will not be simultaneous; the receiver will move on but not at the moment when the generator recoils. If the energy propagates itself indefinitely without encountering a receiver, the compensation will never be made.

Do we say that the space which separates the generator from the receiver and which the perturbation must pass over in going from the one to the other is not void, that it is full not only of ether, but of air; or even in the interplanetary spaces of some fluid subtle but still ponderable; that this matter undergoes the shock like the receiver at the moment when the energy reaches it, and recoils in its turn when the perturbation quits it? That would save the principle of Newton, but that is not true.

If energy in its diffusion remained always attached to some material substratum, then matter in motion would carry along light with it, and Fizeau has demonstrated that it does nothing of the sort, at least for air. This is what Michelson and Morley have since confirmed.

One may suppose also that the movements of matter, properly so called, are exactly compensated by those of the ether; but that would lead us to the same reflections as just now. The principle so extended would explain everything, since whatever might be the visible movements, we should always have the power of imagining hypothetical movements which compensated them.

But if it is able to explain everything, this is because it does not permit us to foresee anything; it does not enable us to decide between different possible hypotheses, since it explains everything beforehand. It therefore becomes useless.

And then the suppositions that it would be necessary to make on the movements of the ether are not very satisfactory.

If the electric charges double, it would be natural to imagine that the velocities of the divers atoms of ether double also, and for the compensation, it would be necessary that the mean velocity of the ether quadruple.

This is why I have long thought that these consequences of theory, contrary to the principle of Newton, would end some day by being abandoned, and yet the recent experiments on the movements of the electrons issuing from radium seem rather to confirm them.

I arrive at the principle of Lavoisier on the conservation of masses: in truth this is one not to be touched without unsettling all mechanics.

And now certain persons believe that it seems true to us only because we consider in mechanics merely moderate velocities, but that it would cease to be true for bodies animated by velocities comparable to that of light. These velocities, it is now believed, have been realized; the cathode rays or those of radium may be formed of very minute particles or of electrons which are displaced with velocities smaller no doubt than that of light, but which might be its one tenth or one third.

These rays can be deflected, whether by an electric field, or by a magnetic field, and we are able by comparing these deflections, to measure at the same time the velocity of the electrons and their mass (or rather the relation of their mass to their charge). But when it was seen that these velocities approached that of light, it was decided that a correction was necessary.

These molecules, being electrified, could not be displaced without agitating the ether; to put them in motion it is necessary to overcome a double inertia, that of the molecule itself and that of the ether. The total or apparent mass that one measures is composed, therefore, of two parts: the real or mechanical mass of the molecule and the electro-dynamic mass representing the inertia of the ether.

The calculations of Abraham and the experiments of Kaufmann have then shown that the mechanical mass, properly so called, is null, and that the mass of the electrons, or, at least, of the negative electrons, is of exclusively electro-dynamic origin. This forces us to change the definition of mass; we cannot any longer distinguish mechanical mass and electro-dynamic mass, since then the first would vanish; there is no mass other than electro-dynamic inertia. But in this case the mass can no longer be constant, it augments with the velocity, and it even depends on the direction, and a body animated by a notable velocity will not oppose the same inertia to the forces which tend to deflect it from its route, as to those which tend to accelerate or to retard its progress.

There is still a resource; the ultimate elements of bodies are electrons, some charged negatively, the others charged positively. The negative electrons have no mass, this is understood; but the positive electrons, from the little we know of them, seem much greater. Perhaps they have, besides their electro-dynamic mass, a true mechanical mass. The veritable mass of a body would, then, be the sum of the mechanical masses of its positive electrons, the negative electrons not counting; mass so defined could still be constant.

Alas, this resource also evades us. Recall what we have said of the principle of relativity and of the efforts made to save it. And it is not merely a principle which it is a question of saving, such are the indubitable results of the experiments of Michelson.

Lorentz has been obliged to suppose that all the forces, whatever be their origin, were affected with a coefficient in a medium animated by a uniform translation; this is not sufficient; it is still necessary, says he, that *the masses of all the particles be influenced by a translation to the same degree as the electro-magnetic masses of the electrons.*

So the mechanical masses will vary in accordance with the same laws as the electro-dynamic masses; they cannot, therefore, be constant.

Need I point out that the fall of the principle of Lavoisier involves that of the principle of Newton? This latter signifies that the centre of gravity of an isolated system moves in a straight line; but if there is no longer a constant mass, there is no longer a centre of gravity, we no longer know even what this is. This is why I said above that the experiments on the cathode rays appeared to justify the doubts of Lorentz on the subject of the principle of Newton.

From all these results, if they are confirmed, would arise an entirely new mechanics, which would be, above all, characterized by this fact, that no velocity could surpass that of light, any more than any temperature could fall below the zero absolute, because bodies would oppose an increasing inertia to the causes, which would tend to accelerate their motion; and this inertia would become infinite when one approached the velocity of light.

Nor for an observer carried along himself in a translation he did not suspect could any apparent velocity surpass that of light; there would then be a contradiction, if we recall that this observer would not use the same clocks as a fixed observer, but, indeed, clocks marking "local time."

Here we are then facing a question I content myself with stating. If there is no longer any mass, what becomes of the law of Newton?

Mass has two aspects, it is at the same time a coefficient of inertia and an attracting mass entering as factor into Newtonian attraction. If the coefficient of inertia is not constant, can the attracting mass be? That is the question. . . .

In presence of this general ruin of the principles, what attitude will mathematical physics take? . . .

Should we not also endeavor to obtain a more satisfactory theory of the electro-dynamics of bodies in motion? It is there especially, as I have sufficiently shown above, that difficulties accumulate. Evidently we must heap up hypotheses, we cannot satisfy all the principles at once; heretofore, one has succeeded in safeguarding some only on condition of sacrificing the others; but all hope of obtaining better results is not yet lost. Let us take, therefore, the theory of Lorentz, turn it in all senses, modify it little by little, and perhaps everything will arrange itself.

Thus in place of supposing that bodies in motion undergo a contraction in the sense of the motion, and that this contraction is the same whatever be the nature of these bodies and the forces to which they are otherwise submitted, could we not make an hypothesis more simple and more natural?

We might imagine, for example, that it is the ether which is modified when it is in relative motion in reference to the material medium which it penetrates, that when it is thus modified, it no longer transmits perturbations with the same velocity in every direction. It might transmit more rapidly those which are propagated parallel to the medium, whether in the same sense or in the opposite sense, and less rapidly those which are propagated perpendicularly. The wave surfaces would no longer be spheres, but ellipsoids, and we could dispense with that extraordinary contraction of all bodies.

I cite that only as an example, since the modifications one might essay would be evidently susceptible of infinite variation. . . .

We cannot foresee in what way we are about to expand; perhaps it is the kinetic theory of gases which is about to undergo development and serve as model to the others. Then, the facts which first appeared to us as simple, thereafter will be merely results of a very great number of elementary facts which only the laws of chance make coöperate for a common end. Physical law will then take an entirely new aspect; it will no longer be solely a differential equation, it will take the character of a statistical law.

Perhaps, likewise, we should construct a whole new mechanics, of which we only succeed in catching a glimpse, where inertia increasing with the velocity, the velocity of light would become an impassable limit.

The ordinary mechanics, more simple, would remain a first approximation, since it would be true for velocities not too great, so that we should still find the old dynamics under the new.

We should not have to regret having believed in the principles, and even, since velocities too great for the old formulas would always be only exceptional, the surest way in practice would be still to act as if we continued to believe in them. They are so useful, it would be necessary to keep a place for them. To determine to exclude them altogether would be to deprive one's self of a precious weapon. I hasten to say in conclusion we are not yet there, and as yet nothing proves that the principles will not come forth from the combat victorious and intact.

6 *Albert Einstein*
"On the Electrodynamics of Moving Bodies"

In 1905 a relatively obscure employee of the Swiss Patent Office at Berne published three papers in the German journal Annalen der Physik. *Each was of fundamental importance in the history of physics. One reduced the confusion of the mechanism of Brownian movement to order and, incidentally, established beyond all reasonable doubt the real existence of molecules. Another applied the new quantum theory to the photoelectric effect in a way that revolutionized the study of physical optics. It was for this work that the author received the Nobel Prize in Physics in 1921. The third dealt with the subject of the electrodynamics of moving bodies, that field which Poincaré had suggested might reveal important new laws. It was in this paper that Albert Einstein (1879–1955) spelled out the outlines of the Special Theory of Relativity. The selection that follows contains only the nonmathematical part of the paper, but it includes all the major assumptions made by Einstein.*

It is known that Maxwell's electrodynamics—as usually understood at the present time—when applied to moving bodies, leads

SOURCE. H. A. Lorentz, A. Einstein, H. Minkowski, and H. Weyl, *The Principal of Relativity*, W. Perrett and G. B. Jeffery, trs., New York: Dover Publications, Inc., 1923, pp. 37–43. Reprinted with the permission of the publisher.

to asymmetries which do not appear to be inherent in the phenomena. Take, for example, the reciprocal electrodynamic action of a magnet and a conductor. The observable phenomenon here depends only on the relative motion of the conductor and the magnet, whereas the customary view draws a sharp distinction between the two cases in which either the one or the other of these bodies is in motion. For if the magnet is in motion and the conductor at rest, there arises in the neighbourhood of the magnet an electric field with a certain definite energy, producing a current at the places where parts of the conductor are situated. But if the magnet is stationary and the conductor in motion, no electric field arises in the neighbourhood of the magnet. In the conductor, however, we find an electromotive force, to which in itself there is no corresponding energy, but which gives rise—assuming equality of relative motion in the two cases discussed—to electric currents of the same path and intensity as those produced by the electric forces in the former case.

Examples of this sort, together with the unsuccessful attempts to discover any motion of the earth relatively to the "light medium," suggest that the phenomena of electrodynamics as well as of mechanics possess no properties corresponding to the idea of absolute rest. They suggest rather that, as has already been shown to the first order of small quantities, the same laws of electrodynamics and optics will be valid for all frames of reference for which the equations of mechanics hold good. We will raise this conjecture (the purport of which will hereafter be called the "Principle of Relativity") to the status of a postulate, and also introduce another postulate, which is only apparently irreconcilable with the former, namely, that light is always propagated in empty space with a definite velocity c which is independent of the state of motion of the emitting body. These two postulates suffice for the attainment of a simple and consistent theory of the electrodynamics of moving bodies based on Maxwell's theory for stationary bodies. The introduction of a "luminiferous ether" will prove to be superfluous inasmuch as the view here to be developed will not require an "absolutely stationary space" provided with special properties, nor assign a velocity-vector to a point of the empty space in which electromagnetic processes take place.

The theory to be developed is based—like all electrodynamics—on the kinematics of the rigid body, since the assertions of any such theory have to do with the relationships between rigid bodies (systems of co-ordinates), clocks, and electromagnetic processes. Insufficient consideration of this circumstance lies at the root of the difficulties which the electrodynamics of moving bodies at present encounters.

I. KINEMATICAL PART

1. Definition of Simultaneity. Let us take a system of co-ordinates in which the equations of Newtonian mechanics hold good. In order to render our presentation more precise and to distinguish this system of co-ordinates verbally from others which will be introduced hereafter, we call it the "stationary system."

If a material point is at rest relatively to this system of co-ordinates, its position can be defined relatively thereto by the employment of rigid standards of measurement and the methods of Euclidean geometry, and can be expressed in Cartesian co-ordinates.

If we wish to describe the *motion* of a material point, we give the values of its co-ordinates as functions of the time. Now we must bear carefully in mind that a mathematical description of this kind has no physical meaning unless we are quite clear as to what we understand by "time." We have to take into account that all our judgments in which time plays a part are always judgments of *simultaneous events.* If, for instance, I say, "That train arrives here at 7 o'clock," I mean something like this: "The pointing of the small hand of my watch to 7 and the arrival of the train are simultaneous events."

It might appear possible to overcome all the difficulties attending the definition of "time" by substituting "the position of the small hand of my watch" for "time." And in fact such a definition is satisfactory when we are concerned with defining a time exclusively for the place where the watch is located; but it is no longer satisfactory when we have to connect in time series of events occurring at different places, or—what comes to the same thing—to evaluate the times of events occurring at places remote from the watch.

We might, of course, content ourselves with time values determined by an observer stationed together with the watch at the origin of the co-ordinates, and co-ordinating the corresponding positions of the hands with light signals, given out by every event to be timed, and reaching him through empty space. But this co-ordination has the disadvantage that it is not independent of the standpoint of the observer with the watch or clock, as we know from experience. We arrive at a much more practical determination along the following line of thought.

If at the point A of space there is a clock, an observer at A can determine the time values of events in the immediate proximity of A by finding the positions of the hands which are simultaneous with these events. If there is at the point B of space another clock in all respects resembling the one at A, it is possible for an observer at B to determine the time values of events in the immediate neighbourhood of B. But it is not possible without further assumption to compare, in respect of time, an event at A with an event at B. We have so far defined only an "A time" and a "B time." We have not defined a common "time" for A and B, for the latter cannot be defined at all unless we establish *by definition* that the"time" required by light to travel from A to B equals the "time" it requires to travel from B to A. Let a ray of light start at the "A time" t_A from A towards B, let it at the "B time" t_B be reflected at B in the direction of A, and arrive again at A at the "A time" t_B'.

In accordance with definition the two clocks synchronize if

$$t_B - t_A = t'_A - t_B.$$

We assume that this definition of synchronism is free from contradictions, and possible for any number of points; and that the following relations are universally valid:

1. If the clock at B synchronizes with the clock at A, the clock at A synchronizes with the clock at B.

2. If the clock at A synchronizes with the clock at B and also with the clock at C, the clocks at B and C also synchronize with each other.

Thus with the help of certain imaginary physical experiments we have settled what is to be understood by synchronous stationary clocks located at different places, and have evidently obtained a definition of "simultaneous," or "synchronous," and of "time." The "time" of an event is that which is given simultaneously with the event by a stationary clock located at the place of the event, this clock being synchronous, and indeed synchronous for all time determinations, with a specified stationary clock.

In agreement with experience we further assume the quantity

$$\frac{2AB}{t'_A - t_A} = c,$$

to be a universal constant—the velocity of light in empty space.

It is essential to have time defined by means of stationary clocks in the stationary system, and the time now defined being appropriate to the stationary system we call it "the time of the stationary system."

2. *On the Relativity of Lengths and Times.* The following reflexions are based on the principle of relativity and on the principle of the constancy of the velocity of light. These two principles we define as follows:

1. The laws by which the states of physical systems undergo change are not affected, whether these changes of state be referred to the one or the other of two system of co-ordinates in uniform translatory motion.

2. Any ray of light moves in the "stationary" system of co-ordinates with the determined velocity c, whether the ray be emitted by a stationary or by a moving body. Hence.

$$\text{velocity} = \frac{\text{light path}}{\text{time interval}}$$

where time interval is to be taken in the sense of the definition in § 1.

Let there be given a stationary rigid rod; and let its length be l as measured by a measuring-rod which is also stationary. We now imagine the axis of the rod lying along the axis of x of the stationary system of co-ordinates, and that a uniform motion of parellel translation with velocity v along the axis of x in the direction of increasing x is then imparted to the rod. We now inquire as to the length of the moving rod, and imagine its length to be ascertained by the following two operations:

(*a*) The observer moves together with the given measuring-rod and the rod to be measured, and measures the length of the rod directly by superposing the measuring-rod, in just the same way as if all three were at rest.

(*b*) By means of stationary clocks set up in the stationary system and synchronizing in accordance with § 1, the observer ascertains at what points of the stationary system the two ends of the rod to be measured are located at a definite time. The distance between these two points, measured by the measuring-rod already employed, which in this case is at rest, is also a length which may be designated "the length of the rod."

In accordance with the principle of relativity the length to be discovered by the operation (*a*)—we will call it "the length of the rod in the moving system"—must be equal to the length l of the stationary rod.

The length to be discovered by the operation (*b*) we will call "the length of the (moving) rod in the stationary system." This we shall determine on the basis of our two principles, and we shall find that it differs from l.

Current kinematics tacitly assumes that the lengths determined by these two operations are precisely equal, or in other words, that a moving rigid body at the epoch t may in geometrical respects be perfectly represented by *the same* body *at rest* in a definite position.

We imagine further that at the two ends A and B of the rod, clocks are placed which synchronize with the clocks of the stationary system, that is to say that their indications correspond at any instant to the "time of the stationary system" at the places where they happen to be. These clocks are therefore "synchronous in the stationary system."

We imagine further that with each clock there is a moving observer, and that these observers apply to both clocks the criterion established in § 1 for the synchronization of two clocks. Let a ray of light depart from A at the time t_A, let it be reflected at B at the time t_B, and reach A again at the time t'_A. Taking into consideration the principle of the constancy of the velocity of light we find that

$$t_B - t_A = \frac{r_{AB}}{c - v} \text{ and } t'_A - t_B = \frac{r_{AB}}{c + v}$$

where r_{AB} denotes the length of the moving rod–measured in the stationary system. Observers moving with the moving rod would thus find that the two clocks were not synchronous, while observers in the stationary system would declare the clocks to be synchronous.

So we see that we cannot attach any *absolute* signification to the concept of simultaneity, but that two events which, veiwed from a system of co-ordinates, are simultaneous, can no longer be looked upon as simultaneous events when envisaged from a system which is in motion relatively to that system.

PART II

The Nature of Relativity Theory

1 FROM *Albert Einstein* *Relativity, The Special and General Theory*

The literature on Relativity Theory is vast. Literally thousands of books or articles have been written to explain what relativity is all about to the layman who cannot follow the mathematical equations with which all is made clear to the physicist. One of the best of these popularizations was written by Einstein himself.

SPACE AND TIME IN CLASSICAL MECHANICS

"The purpose of mechanics is to describe how bodies change their position in space with time." I should load my conscience with grave sins against the sacred spirit of lucidity were I to formulate the aims of mechanics in this way, without serious reflection and detailed explanations. Let us proceed to disclose these sins.

It is not clear what is to be understood here by "position" and "space." I stand at the window of a railway carriage which is travelling uniformly, and drop a stone on the embankment, without throwing it. Then, disregarding the influence of the air resistance, I see the stone descend in a straight line. A pedestrian who observes the misdeed from the footpath notices that the stone falls to earth in a parabolic curve. I now ask: Do the "positions" traversed by the stone lie "in reality" on a straight line or on a parabola? Moreover, what is meant here by motion "in space"? From the considerations of the previous section the answer is self-evident. In the first place, we entirely shun the vague word "space," of which, we must honestly acknowledge, we cannot form the slightest conception, and we replace it by "motion relative to a practically rigid body of reference." The posi-

SOURCE. Albert Einstein, *Relativity, The Special and General Teory*, Robert W. Lawson, tr., New York: Henry Holt and Company, 1921, pp. 9–11, 14–18, 21–33, 52–57, 74–81, and 135-137. Reprinted by permission of Peter Smith, Publisher.

tions relative to the body of reference (railway carriage or embankment) have already been defined in detail in the preceding section. If instead of "body of reference" we insert "system of co-ordinates," which is a useful idea for mathematical description, we are in a position to say: The stone traverses a straight line relative to a system of co-ordinates rigidly attached to the carriage, but relative to a system of co-ordinates rigidly attached to the ground (embankment) it describes a parabola. With the aid of this example it is clearly seen that there is no such thing as an independently existing trajectory (lit. "path-curve"), but only a trajectory relative to a particular body of reference.

In order to have a *complete* description of the motion, we must specify how the body alters its position *with time; i.e.*, for every point on the trajectory it must be stated at what time the body is situated there. These data must be supplemented by such a definition of time that, in virtue of this definition, these time-values can be regarded essentially as magnitudes (results of measurements) capable of observation. If we take our stand on the ground of classical mechanics, we can satisfy this requirement for our illustration in the following manner. We imagine two clocks of identical construction; the man at the railway-carriage window is holding one of them, and the man on the footpath the other. Each of the observers determines the position on his own reference-body occupied by the stone at each tick of the clock he is holding in his hand. In this connection we have not taken account of the inaccuracy involved by the finiteness of the velocity of propagation of light. With this and with a second difficulty prevailing here we shall have to deal in detail later. . . .

THE PRINCIPLE OF RELATIVITY (IN THE RESTRICTED SENSE)

In order to attain the greatest possible clearness, let us return to our example of the railway carriage supposed to be travelling uniformly. We call its motion a uniform translation ("uniform" because it is of constant velocity and direction, "translation" because although the carriage changes its position relative to the embankment yet it does not rotate in so doing). Let us imagine a raven flying through the air in such a manner that its motion, as observed from the embankment, is uniform and in a straight line. If we were to observe the flying raven from the moving railway carriage, we should find that the motion of the raven would be one of different velocity and direction, but that it would still be uniform and in a straight line. Expressed in an abstract manner we may say: If a mass m is moving uniformly in a straight line with respect to a co-ordinate system K, then it will also be moving uniformly and in a straight line relative to a second co-ordinate system K', provided that the latter is executing a uniform translatory motion with respect to K. In accordance with the discussion contained in the preceding section, it follows that:

If K is a Galileian co-ordinate system,* then every other co-ordinate system K' is a Galileian one, when, in relation to K, it is in a condition of uniform motion of translation. Relative to K' the mechanical laws of Galilei-Newton hold good exactly as they do with respect to K.

We advance a step farther in our generalisation when we express the tenet thus: If, relative to K, K' is a uniformly moving co-ordinate system devoid of rotation, then natural phenomena run their course with respect to K' according to exactly the same general laws as with respect to K. This statement is called the *principle of relativity* (in the restricted sense). . . .

We now proceed to the second argument, to which, moreover, we shall return later. If the principle of relativity (in the restricted sense) does not hold, then the Galileian co-ordinate systems K, K', K'', etc., which are moving uniformly relative to each other, will not be *equivalent* for the description of natural phenomena. In this case we should be constrained to believe that natural laws are capable of being formulated in a particularly simple manner, and of course only on condition that, from amongst all possible Galileian co-ordinate systems, we should

* That is, the coordinate system of classical mechanics. If a man A moving in the X direction relative to an observer B at velocity V_1 threw a stone forward at velocity V_2, the velocity in A's frame of reference (V_2) would be transformed into the velocity in B's frame of reference by adding to it A's velocity (V_1). This is a Galileian transformation [Ed.].

have chosen *one* (K_0) of a particular state of motion as our body of reference. We should then be justified (because of its merits for the description of natural phenomena) in calling this system "absolutely at rest," and all other Galileian systems K "in motion." If, for instance, our embankment were the system K_0, then our railway carriage would be a system K, relative to which less simple laws would hold than with respect to K_0. This diminished simplicity would be due to the fact that the carriage K would be in motion (*i.e.*, "really") with respect to K_0. In the general laws of nature which have been formulated with reference to K, the magnitude and direction of the velocity of the carriage would necessarily play a part. We should expect, for instance, that the note emitted by an organ-pipe placed with its axis parallel to the direction of travel would be different from that emitted if the axis of the pipe were placed perpendicular to this direction. Now in virtue of its motion in an orbit round the sun, our earth is comparable with a railway carriage travelling with a velocity of about 30 kilometres per second. If the principle of relativity were not valid we should therefore expect that the direction of motion of the earth at any moment would enter into the laws of nature, and also that physical systems in their behaviour would be dependent on the orientation in space with respect to the earth. For owing to the alteration in direction of the velocity of revolution of the earth in the course of a year, the earth cannot be at rest relative to the hypothetical system K_0 throughout the whole year. However, the most careful observations have never revealed such anisotropic properties in terrestrial physical space, *i.e.* a physical non-equivalence of different directions. This is a very powerful argument in favour of the principle of relativity. . . .

THE APPARENT INCOMPATIBILITY OF THE LAW OF PROPAGATION OF LIGHT WITH THE PRINCIPLE OF RELATIVITY

There is hardly a simpler law in physics than that according to which light is propagated in empty space. Every child at school knows, or believes he knows, that this propagation takes place in straight lines with a velocity $c = 300{,}000$ km./sec. At all events we know with great exactness that this velocity is the same for all colours, because if this were not the case, the minimum of emission would not be observed simultaneously for different colours during the eclipse of a fixed star by its dark neighbour. By means of similar considerations based on observations of double stars, the Dutch astronomer De Sitter was also able to show that the velocity of propagation of light cannot depend on the velocity of motion of the body emitting the light. The assumption that this velocity of propagation is dependent on the direction "in space" is in itself improbable.

In short, let us assume that the simple law of the constancy of the velocity of light c (in vacuum) is justifiably believed by the child at school. Who would imagine that this simple law has plunged the conscientiously thoughtful physicist into the greatest intellectual difficulties? Let us consider how these difficulties arise.

Of course we must refer the process of the propagation of light (and indeed every other process) to a rigid reference-body (co-ordinate system). As such a system let us again choose our embankment. We shall imagine the air above it to have been removed. If a ray of light be sent along the embankment, we see from the above that the tip of the ray will be transmitted with the velocity c relative to the embankment. Now let us suppose that our railway carriage is again travelling along the railway lines with the velocity v, and that its direction is the same as that of the ray of light, but its velocity of course much less. Let us inquire about the velocity of propagation of the ray of light relative to the carriage. It is obvious that we can here apply the consideration of the previous section, since the ray of light plays the part of the man walking along relatively to the carriage. The velocity W of the man relative to the embankment is here replaced by the velocity of light relative to the embankment. w is the required velocity of light with respect to the carriage, and we have

$$w = c - v.$$

The velocity of propagation of a ray of light relative to the carriage thus comes out smaller than c.

But this result comes into conflict with the principle of relativity set forth in Section V [*On the Electrodynamics of Moving Bodies*]. For, like every other general law of nature, the law of the transmission of light *in vacuo* must, according to the principle of relativity, be the same for the railway carriage as reference-body as when the rails are the body of reference. But, from our above consideration, this would appear to be impossible. If every ray of light is propagated relative to the embankment with the velocity *c*, then for this reason it would appear that another law of propagation of light must necessarily hold with respect to the carriage—a result contradictory to the principle of relativity.

In view of this delemma there appears to be nothing else for it than to abandon either the principle of relativity or the simple law of the propagation of light *in vacuo*. Those of you who have carefully followed the preceding discussion are almost sure to expect that we should retain the principle of relativity, which appeals so convincingly to the intellect because it is so natural and simple. The law of the propagation of light *in vacuo* would then have to be replaced by a more complicated law conformable to the principle of relativity. The development of theoretical physics shows, however, that we cannot pursue this course. The epoch-making theoretical investigations of H. A. Lorentz on the electrodynamical and optical phenomena connected with moving bodies show that experience in this domain leads conclusively to a theory of electromagnetic phenomena, of which the law of the constancy of the velocity of light *in vacuo* is a necessary consequence. Prominent theoretical physicists were therefore more inclined to reject the principle of relativity, in spite of the fact that no empirical data had been found which were contradictory to this principle.

At this juncture the theory of relativity entered the arena. As a result of an analysis of the physical conceptions of time and space, it became evident that *in reality there is not the least incompatibility between the principle of relativity and the law of propagation of light*, and that by systematically holding fast to both these laws a logically rigid theory could be arrived at. This theory has been called the *special theory of relativity* to distinguish it from the extended theory, with which we shall deal later. In the following pages we shall present the fundamental ideas of the special theory of relativity.

ON THE IDEA OF TIME IN PHYSICS

Lightning has struck the rails on our railway embankment at two places *A* and *B* far distant from each other. I make the additional assertion that these two lightning flashes occurred simultaneously. If I ask you whether there is sense in this statement, you will answer my question with a decided "Yes." But if I now approach you with the request to explain to me the sense of the statement more precisely, you find after some consideration that the answer to this question is not so easy as it appears at first sight.

After some time perhaps the following answer would occur to you: "The significance of the statement is clear in itself and needs no further explanation; of course it would require some consideration if I were to be commissioned to determine by observations whether in the actual case the two events took place simultaneously or not." I cannot be satisfied with this answer for the following reason. Supposing that as a result of ingenious considerations an able meteorologist were to discover that the lightning must always strike the places *A* and *B* simultaneously, then we should be faced with the task of testing whether or not this theoretical result is in accordance with the reality. We encounter the same difficulty with all physical statements in which the conception "simultaneous" plays a part. The concept does not exist for the physicist until he has the possibility of discovering whether or not it is fulfilled in an actual case. We thus require a definition of simultaneity such that this definition supplies us with the method by means of which, in the present case, he can decide by experiment whether or not both the lightning strokes occurred simultaneously. As long as this requirement is not satisfied, I allow myself to be deceived as a physicist (and of course the same applies if I am not a physicist), when I imagine that I am able to attach a meaning to the statement of simultaneity. (I would ask the reader not to proceed farther until he is fully convinced on this point.)

After thinking the matter over for some time you then offer the following suggestion with which to test simultaneity. By measuring along the rails, the connecting line *AB* should be measured up and an observer placed at the mid-point *M* of the distance AB. This observer should be supplied with an arrangement (*e.g.* two mirrors inclined at 90°) which allows him visually to observe both places A and B at the same time. If the observer perceives the two flashes of lightning at the same time, then they are simultaneous.

I am very pleased with this suggestion, but for all that I cannot regard the matter as quite settled, because I feel constrained to raise the following objection: "Your definition would certainly be right, if I only knew that the light by means of which the observer at M perceives the lightning flashes travels along the length $A \longrightarrow M$ with the same velocity as along the length $B \longrightarrow M$. But an examination of this supposition would only be possible if we already had at our disposal the means of measuring time. It would thus appear as though we were moving here in a logical circle."

After further consideration you cast a somewhat disdainful glance at me–and rightly so–and you declare: "I maintain my previous definition nevertheless, because in reality it assumes absolutely nothing about light. There is only *one* demand to be made of the definition of simultaneity, namely, that in every real case it must supply us with an empirical decision as to whether or not the conception that has to be defined is fulfilled. That my definition satisfies this demand is indisputable. That light requires the same time to traverse the path $A \longrightarrow M$ as for the path $B \longrightarrow M$ is in reality neither a *supposition nor a hypothesis* about the physical nature of light, but a *stipulation* which I can make of my own freewill in order to arrive at a definition of simultaneity."

It is clear that this definition can be used to give an exact meaning not only to *two* events, but to as many events as we care to choose, and independently of the positions of the scenes of the events with respect to the body of reference (here the railway embankment). We are thus led also to a definition of "time" in physics. For this purpose we suppose that clocks of identical construction are placed at the points A, B and C of the railway line (co-ordinate system), and that they are set in such a manner that the positions of their pointers are simultaneously (in the above sense) the same. Under these conditions we understand by the "time" of an event the reading (position of the hands) of that one of these clocks which is in the immediate vicinity (in space) of the event. In this manner a time-value is associated with every event which is essentially capable of observation.

This stipulation contains a further physical hypothesis, the validity of which will hardly be doubted without empirical evidence to the contrary. It has been assumed that all these clocks go *at the same rate* if they are of identical construction. Stated more exactly: When two clocks arranged at rest in different places of a reference-body are set in such a manner that a *particular* position of the pointers of the one clock is *simultaneous* (in the above sense) with the *same* position of the pointers of the other clock, then identical "settings" are always simultaneous (in the sense of the above definition).

THE RELATIVITY OF SIMULTANEITY

Up to now our considerations have been referred to a particular body of reference, which we have styled a "railway embankment." We suppose a very long train travelling along the rails with the constant velocity v and in the direction indicated in Fig. 1. People travelling in this train will with advantage use the train as a rigid reference-body (co-ordinate system); they

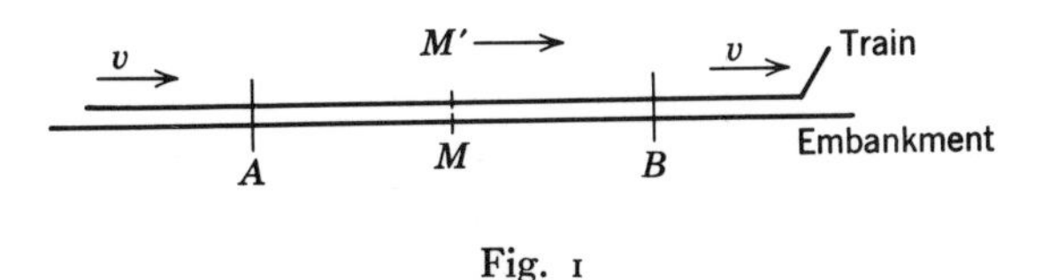

Fig. 1

regard all events in reference to the train. Then every event which takes place along the line also takes place at a particular point of the train. Also the definition of simultaneity can be given relative to the train in exactly the same way as with respect to the embankment. As a natural consequence, however, the following question arises:

Are two events (*e.g.* the two strokes of lightning A and B)

which are simultaneous *with reference to the railway embankment* also simultaneous *relatively to the train?* We shall show directly that the answer must be in the negative.

When we say that the lightning strokes A and B are simultaneous with respect to the embankment, we mean: the rays of light emitted at the places A and B, where the lightning occurs, meet each other at the mid-point M of the length $A \longrightarrow B$ of the embankment. But the events A and B also correspond to positions A and B on the train. Let M' be the mid-point of the distance $A \longrightarrow B$ on the travelling train. Just when the flashes of lightning occur, this point M' naturally coincides with the point M, but it moves towards the right in the diagram with the velocity v of the train. If an observer sitting in the position M' in the train did not possess this velocity, then he would remain permanently at M, and the light rays emitted by the flashes of lightning A and B would reach him simultaneously, *i.e.* they would meet just where he is situated. Now in reality (considered with reference to the railway embankment) he is hastening towards the beam of light coming from B, whilst he is riding on ahead of the beam of light coming from A. Hence the observer will see the beam of light emitted from B earlier than he will see that emitted from A. Observers who take the railway train as their reference-body must therefore come to the conclusion that the lightning flash B took place earlier than the lightning flash A. We thus arrive at the important result:

Events which are simultaneous with reference to the embankment are not simultaneous with respect to the train, and *vice versa* (relativity of simultaneity). Every reference-body (coordinate system) has its own particular time; unless we are told the reference-body to which the statement of time refers, there is no meaning in a statement of the time of an event.

Now before the advent of the theory of relativity it had always tacitly been assumed in physics that the statement of time had an absolute significance, *i.e.* that it is independent of the state of motion of the body of reference. But we have just seen that this assumption is incompatible with the most natural definition of simultaneity; if we discard this assumption, then the conflict between the law of the propagation of light *in vacuo* and the principle of relativity (developed in Section VII) disappears.

We were led to that conflict by the considerations of Section VI, which are now no longer tenable. In that section we concluded that the man in the carriage, who traverses the distance *w per second* relative to the carriage, traverses the same distance also with respect to the embankment *in each second* of time. But, according to the foregoing considerations, the time required by a particular occurrence with respect to the carriage must not be considered equal to the duration of the same occurrence as judged from the embankment (as reference-body). Hence it cannot be contended that the man in walking travels the distance w relative to the railway line in a time which is equal to one second as judged from the embankment. . . .

THE LORENTZ TRANSFORMATION

The results of the last three sections show that the apparent incompatibility of the law of propagation of light with the principle of relativity (Section VII) has been derived by means of a consideration which borrowed two unjustifiable hypotheses from classical mechanics; these are as follows:

1. The time-interval (time) between two events is independent of the condition of motion of the body of reference.
2. The space-interval (distance) between two points of a rigid body is independent of the condition of motion of the body of reference.

If we drop these hypotheses, then the dilemma of Section VII disappears, because the theorem of the addition of velocities derived in Section VI becomes invalid. The possibility presents itself that the law of the propagation of light *in vacuo* may be compatible with the principle of relativity, and the question arises: How have we to modify the considerations of Section VI in order to remove the apparent disagreement between these two fundamental results of experience? This question leads to a general one. In the discussion of Section VI we have to do with places and times relative both to the train and to the embankment. How are we to find the place and time of an event in relation to the train, when we know the place and time of the event with respect to the railway embankment? Is there a thinkable answer to this question of such a nature that the law of transmission of light *in vacuo* does not contradict the principle of relativity? In other words: Can we conceive of a relation between place and time of the individual events relative to both reference-bodies, such that every ray of light possesses the velocity of transmission c relative to the embankment and relative to the train? This question leads to a quite definite positive answer, and to a perfectly definite transformation law for the space-time magnitudes of an event when changing over from one body of reference to another.

Before we deal with this, we shall introduce the following incidental consideration. Up to the present we have only considered events taking place along the embankment, which had mathematically to assume the function of a straight line. In the manner indicated in Section II we can imagine this reference-body supplemented laterally and in a vertical direction by means of a framework of rods, so that an event which takes place anywhere can be localised with reference to this framework. Similarly, we can imagine the train travelling with the velocity v to be continued across the whole of space, so that every event, no matter how far off it may be, could also be localised with respect to the second framework. Without committing any fundamental error, we can disregard the fact that in reality these frameworks would continually interfere with each other, owing to the impenetrability of solid bodies. In every such framework we imagine three surfaces perpendicular to each other marked out, and designated as "co-ordinate planes" ("co-ordinate system"). A co-ordinate system K then corresponds to the embankment, and a co-ordinate system K' to the train. An event, wherever it may have taken place, would be fixed in space with respect to K by the three perpendiculars x, y, z on the co-ordinate planes, and with regard to time by a time-value t. Relative to K', *the same event* would be fixed in respect of space and time by corresponding values x', y', z', t', which of course are not identical with x, y, z, t. It has already been set forth in detail how these magnitudes are to be regarded as results of physical measurements.

Obviously our problem can be exactly formulated in the following manner. What are the values x', y', z', t' of an event with respect to K', when the magnitudes x, y, z, t, of the same event with respect to K are given? The relations must be so chosen that

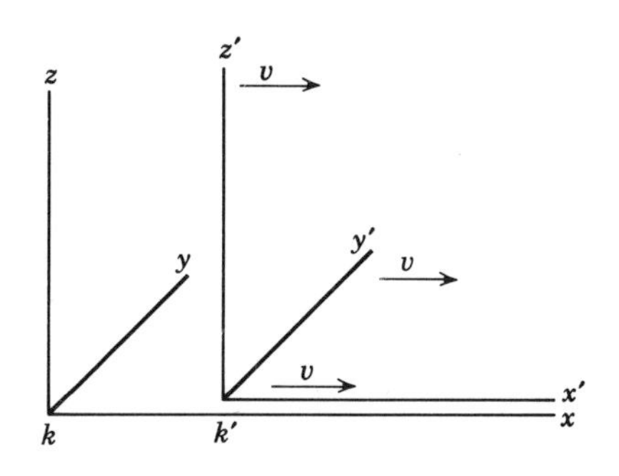

Fig. 2

the law of the transmission of light *in vacuo* is satisfied for one and the same ray of light (and of course for every ray) with respect to K and K'. For the relative orientation in space of the co-ordinate systems indicated in the diagram (Fig. 2), this problem is solved by means of the equations:

$$x' = \frac{x - vt}{\sqrt{1 - \frac{v^2}{c^2}}}$$

$$y' = y$$

$$z' = z$$

$$t' = \frac{t - \frac{v}{c^2}.x}{\sqrt{1 - \frac{v^2}{c^2}}}.$$

This system of equations is known as the "Lorentz transformation."

If in place of the law of transmission of light we had taken as our basis the tacit assumptions of the older mechanics as to the absolute character of times and lengths, then instead of the above we should have obtained the following equations:

$$x' = x - vt$$
$$y' = y$$
$$z' = z$$
$$t' = t.$$

This system of equations is often termed the "Galilei transformation." The Galilei transformation can be obtained from the Lorentz transformation by substituting an infinitely large value for the velocity of light c in the latter transformation.

Aided by the following illustration, we can readily see that, in accordance with the Lorentz transformation, the law of the transmission of light *in vacuo* is satisfied both for the reference-body K and for the reference-body K'. A light-signal is sent along the positive x-axis, and this light-stimulus advances in accordance with the equation

$$x = ct,$$

i.e., with the velocity c. According to the equations of the Lorentz transformation, this simple relation between x and t involves a relation between x' and t'. In point of fact, if we substitute for x the value ct in the first and fourth equations of the Lorentz transformation, we obtain:

$$x' = \frac{(c - v)t}{\sqrt{1 - \frac{v^2}{c^2}}}$$

$$t' = \frac{\left(1 - \frac{v}{c}\right)t}{\sqrt{1 - \frac{v^2}{c^2}}},$$

from which, by division, the expression

$$x' = ct'$$

immediately follows. If referred to the system K', the propagation of light takes place according to this equation. We thus see that the velocity of transmission relative to the reference-body K' is also equal to c. The same result is obtained for rays of light advancing in any other direction whatsoever. Of course this is not surprising, since the equations of the Lorentz transformation were derived conformably to this point of view. . . .

GENERAL RESULTS OF THE THEORY

It is clear from our previous considerations that the (special) theory of relativity has grown out of electrodynamics and optics. In these fields it has not appreciably altered the predictions of theory, but it has considerably simplified the theoretical structure, *i.e.*, the derivation of laws, and—what is incomparably more important—it has considerably reduced the number of independent hypotheses forming the basis of theory. The special theory of relativity has rendered the Maxwell-Lorentz theory so plausible, that the latter would have been generally accepted by physicists even if experiment had decided less unequivocally in its favour.

Classical mechanics required to be modified before it could come into line with the demands of the special theory of relativity. For the main part, however, this modification affects only the laws for rapid motions, in which the velocities of matter v are not very small as compared with the velocity of light. We have experience of such rapid motions only in the case of electrons and ions; for other motions the variations from the laws of classical mechanics are too small to make themselves evident in practice. We shall not consider the motion of stars until we come to speak of the general theory of relativity. In accordance with the theory of relativity the kinetic energy of a material point of mass m is no longer given by the well-known expression

$$m\frac{v^2}{2},$$

but by the expression

$$\frac{mc^2}{\sqrt{1-\frac{v^2}{c^2}}}$$

This expression approaches infinity as the velocity v approaches the velocity of light c. The velocity must therefore always remain less than c, however great may be the energies used to produce the acceleration. If we develop the expression for the kinetic energy in the form of a series, we obtain

$$mc^2 + m\frac{v^2}{2} + \frac{3}{8}m\frac{v^4}{c^2} + \ldots\ldots$$

When v^2/c^2 is small compared with unity, the third of these terms is always small in comparison with the second, which last is alone considered in classical mechanics. The first term mc^2 does not contain the velocity, and requires no consideration if we are only dealing with the question as to how the energy of a point-mass depends on the velocity. We shall speak of its essential significance later.

The most important result of a general character to which the special theory of relativity has led is concerned with the conception of mass. Before the advent of relativity, physics recognised two conservation laws of fundamental importance, namely, the law of the conservation of energy and the law of the conservation of mass; these two fundamental laws appeared to be quite independent of each other. By means of the theory of relativity they have been united into one law. We shall now briefly consider how this unification came about, and what meaning is to be attached to it.

The principle of relativity requires that the law of the conservation of energy should hold not only with reference to a co-ordinate system K, but also with respect to every co-ordinate system K' which is in a state of uniform motion of translation relative to K, or, briefly, relative to every "Galileian" system of co-ordinates. In contrast to classical mechanics, the Lorentz transformation is the deciding factor in the transition from one such system to another.

By means of comparatively simple considerations we are led to draw the following conclusion from these premises, in conjunction with the fundamental equations of the electrodynamics of Maxwell: A body moving with the velocity v, which absorbs an amount of energy E_0 in the form of radiation without suffering an alteration in velocity in the process, has, as a consequence, its energy increased by an amount

$$\frac{E_0}{\sqrt{1-\frac{v^2}{c^2}}}.$$

In consideration of the expression given above for the kinetic energy of the body, the required energy of the body comes out to be

$$\frac{\left(m+\frac{E_0}{c^2}\right)c^2}{\sqrt{1-\frac{v^2}{c^2}}}.$$

Thus the body has the same energy as a body of mass $(m+E_0/c^2)$ moving with the velocity v. Hence we can say: If a body takes up an amount of energy E_0, then its inertial mass increases by an amount E_0/c^2; the inertial mass of a body is not a constant, but varies according to the change in the energy of the body. The inertial mass of a system of bodies can even be regarded as a measure of its energy. The law of the conservation of the mass of a system becomes identical with the law of the conservation of energy, and is only valid provided that the system neither takes up nor sends out energy. Writing the expression for the energy in the form

$$\frac{mc^2+E_0}{\sqrt{1-\frac{v^2}{c^2}}},$$

we see that the term mc^2, which has hitherto attracted our attention, is nothing else than the energy possessed by the body before it absorbed the energy E_0.

A direct comparison of this relation with experiment is not possible at the present time, owing to the fact that the changes in energy E_0 to which we can subject a system are not large enough to make themselves perceptible as a change in the inertial mass of the system. E_0/c^2 is too small in comparison with the mass m, which was present before the alteration of the energy. It is owing to this circumstance that classical mechanics was able to establish successfully the conservation of mass as a law of independent validity.

Let me add a final remark of a fundamental nature. The success of the Faraday-Maxwell interpretation of electromagnetic action at a distance resulted in physicists becoming convinced that there are no such things as instantaneous actions at a distance (not involving an intermediary medium) of the type of Newton's law of gravitation. According to the theory of relativity, action at a distance with the velocity of light always takes the place of instantaneous action at a distance or of action at a distance with an infinite velocity of transmission. This is connected with the fact that the velocity c plays a fundamental rôle in this theory. In Part II we shall see in what way this result becomes modified in the general theory of relativity. . . .

THE GRAVITATIONAL FIELD

"If we pick up a stone and then let it go, why does it fall to the ground?" The usual answer to this question is: "Because it is attracted by the earth." Modern physics formulates the answer rather differently for the following reason. As a result of the more careful study of electromagnetic phenomena, we have come to regard action at a distance as a process impossible without the intervention of some intermediary medium. If, for instance, a magnet attracts a piece of iron, we cannot be content to regard this as meaning that the magnet acts directly on the iron through the intermediate empty space, but we are constrained to imagine —after the manner of Faraday—that the magnet always calls into being something physically real in the space around it, that something being what we call a "magnetic field." In its turn this magnetic field operates on the piece of iron, so that the latter strives to move towards the magnet. We shall not discuss here the justification for this incidental conception, which is indeed a somewhat arbitrary one. We shall only mention that with its aid electromagnetic phenomena can be theoretically represented much more satisfactorily than without it, and this applies particularly to the transmission of electromagnetic waves. The effects of gravitation also are regarded in an analogous manner.

The action of the earth on the stone takes place indirectly. The earth produces in its surroundings a gravitational field, which acts on the stone and produces its motion of fall. As we know from experience, the intensity of the action on a body diminishes according to a quite definite law, as we proceed farther and farther away from the earth. From our point of view this means: The law governing the properties of the gravitational field in space must be a perfectly definite one, in order correctly to represent the diminution of gravitational action with the distance from operative bodies. It is something like this: The body (*e.g.* the earth) produces a field in its immediate neighbourhood directly; the intensity and direction of the field at points farther removed from the body are thence determined by the law which governs the properties in space of the gravitational fields themselves.

In contrast to electric and magnetic fields, the gravitational field exhibits a most remarkable property, which is of fundamental importance for what follows. Bodies which are moving under the sole influence of a gravitational field receive an acceleration, *which does not in the least depend either on the material or on the physical state of the body*. For instance, a piece of lead and a piece of wood fall in exactly the same manner in a gravitational field (*in vacuo*), when they start off from rest or with the same initial velocity. This law, which holds most accurately, can be expressed in a different form in the light of the following consideration.

According to Newton's law of motion, we have

$$\text{(Force)} = \text{(inertial mass)} \times \text{(acceleration)},$$

where the "inertial mass" is a characteristic constant of the accelerated body. If now gravitation is the cause of the acceleration, we then have

$$\text{(Force)} = \text{(gravitational mass)} \times \text{(intensity of the gravitational field)},$$

where the "gravitational mass" is likewise a characteristic constant for the body. From these two relations follows:

$$\text{(acceleration)} = \frac{\text{(gravitational mass)}}{\text{(inertial mass)}} \times \text{(intensity of the gravitational field)}.$$

If now, as we find from experience, the acceleration is to be independent of the nature and the condition of the body and always the same for a given gravitational field, then the ratio of the gravitational to the inertial mass must likewise be the same for all bodies. By a suitable choice of units we can thus make this ratio equal to unity. We then have the following law: The *gravitational* mass of a body is equal to its *inertial* mass.

It is true that this important law had hitherto been recorded in mechanics, but it had not been *interpreted*. A satisfactory interpretation can be obtained only if we recognise the following fact: *The same* quality of a body manifests itself according to circumstances as "inertia" or as "weight" (lit. "heaviness"). In the following section we shall show to what extent this is actually the case, and how this question is connected with the general postulate of relativity.

THE EQUALITY OF INERTIAL AND GRAVITATIONAL MASS AS AN ARGUMENT FOR THE GENERAL POSTULATE OF RELATIVITY

We imagine a large portion of empty space, so far removed from stars and other appreciable masses that we have before us approximately the conditions required by the fundamental law of Galilei. It is then possible to choose a Galileian reference-body for this part of space (world), relative to which points at rest remain at rest and points in motion continue permanently in uniform rectilinear motion. As reference-body let us imagine a spacious chest resembling a room with an observer inside who is equipped with apparatus. Gravitation naturally does not exist for this observer. He must fasten himself with strings to the floor, otherwise the slightest impact against the floor will cause him to rise slowly towards the ceiling of the room.

To the middle of the lid of the chest is fixed externally a hook with rope attached, and now a "being" (what kind of a being is immaterial to us) begins pulling at this with a constant force. The chest together with the observer then begin to move "upwards" with a uniformly accelerated motion. In course of time their velocity will reach unheard-of values—provided that we are viewing all this from another reference-body which is not being pulled with a rope.

But how does the man in the chest regard the process? The acceleration of the chest will be transmitted to him by the reaction of the floor of the chest. He must therefore take up this pressure by means of his legs if he does not wish to be laid out full length on the floor. He is then standing in the chest in exactly the same way as anyone stands in a room of a house on our earth. If he release a body which he previously had in his hand, the acceleration of the chest will no longer be transmitted to this body, and for this reason the body will approach the floor of the chest with an accelerated relative motion. The observer will further convince himself *that the acceleration of the body towards the floor of the chest is always of the same magnitude, whatever kind of body he may happen to use for the experiment.*

Relying on his knowledge of the gravitational field (as it was discussed in the preceding section), the man in the chest will thus come to the conclusion that he and the chest are in a gravitational field which is constant with regard to time. Of course he will be puzzled for a moment as to why the chest does not fall in this gravitational field. Just then, however, he discovers the hook in the middle of the lid of the chest and the rope which is attached to it, and he consequently comes to the conclusion that the chest is suspended at rest in the gravitational field.

Ought we to smile at the man and say that he errs in his conclusion? I do not believe we ought to if we wish to remain consistent; we must rather admit that his mode of grasping the situation violates neither reason nor known mechanical laws. Even though it is being accelerated with respect to the "Galileian space" first considered, we can nevertheless regard the chest as being at rest. We have thus good grounds for extending the principle of relativity to include bodies of reference which are accelerated with respect to each other, and as a result we have gained a powerful argument for a generalised postulate of relativity.

We must note carefully that the possibility of this mode of interpretation rests on the fundamental property of the gravitational field of giving all bodies the same acceleration, or, what comes to the same thing, on the law of the equality of inertial and gravitational mass. If this natural law did not exist, the man in the accelerated chest would not be able to interpret the behaviour of the bodies around him on the supposition of a gravitational field, and he would not be justified on the grounds of experience in supposing his reference-body to be "at rest."

Suppose that the man in the chest fixes a rope to the inner side of the lid, and that he attaches a body to the free end of the rope. The result of this will be to stretch the rope so that it will hang "vertically" downwards. If we ask for an opinion of the cause of tension in the rope, the man in the chest will say: "The suspended body experiences a downward force in the gravitational field, and this is neutralised by the tension of the rope; what determines the magnitude of the tension of the rope is the *gravitational mass* of the suspended body." On the other hand, an observer who is poised freely in space will interpret the condition of things thus: "The rope must perforce take part in the accelerated motion of the chest, and it transmits this motion to the body attached to it. The tension of the rope is just large enough to effect the acceleration of the body. That which determines the magnitude of the tension of the rope is the *inertial mass* of the body." Guided by this example, we see that our extension of the principle of relativity implies the *necessity* of the law of the equality of inertial and gravitational mass. Thus we have obtained a physical interpretation of this law. . . .

THE STRUCTURE OF SPACE ACCORDING TO THE GENERAL THEORY OF RELATIVITY

According to the general theory of relativity, the geometrical properties of space are not independent, but they are determined by matter. Thus we can draw conclusions about the geometrical structure of the universe only if we base our considerations on the state of the matter as being something that is known. We know from experience that, for a suitably chosen co-ordinate system, the velocities of the stars are small as compared with the velocity of transmission of light. We can thus as a rough approximation arrive at a conclusion as to the nature of the universe as a whole, if we treat the matter as being at rest.

We already know from our previous discussion that the behaviour of measuring-rods and clocks is influenced by gravitational fields, *i.e.* by the distribution of matter. This in itself is sufficient to exclude the possibility of the exact validity of Euclidean geometry in our universe. But it is conceivable that our universe differs only slightly from a Euclidean one, and this notion seems all the more probable, since calculations show that the metrics of surrounding space is influenced only to an exceedingly small extent by masses even of the magnitude of our sun. We might imagine that, as regards geometry, our universe behaves analogously to a surface which is irregularly curved in its individual parts, but which nowhere departs appreciably from a plane: something like the rippled surface of a lake. Such a universe might fittingly be called a quasi-Euclidean universe. As regards its space it would be infinite. But calculation shows that in a quasi-Euclidean universe the average density of matter would necessarily be *nil.* Thus such a universe could not be inhabited by matter everywhere. . . .

If we are to have in the universe an average density of matter which differs from zero, however small may be that difference, then the universe cannot be quasi-Euclidean. On the contrary, the results of calculation indicate that if matter be distributed uniformly, the universe would necessarily be spherical (or elliptical). Since in reality the detailed distribution of matter is not uniform, the real universe will deviate in individual parts from the spherical, *i.e.* the universe will be quasi-spherical. But it will be necessarily finite. In fact, the theory supplies us with a simple connection between the space-expanse of the universe and the average density of matter in it.

PART III

How Was Relativity Theory Born?

In the first section it was hinted that the road to the formulation of Relativity Theory might not be so smooth and straight as one might assume. There is, in fact, a controversy today over precisely this point. Some of the opposing points of view are offered in the selections that follow.

1 Albert Einstein
"Autobiographical Notes"

We are fortunate in having Albert Einstein's own account of his intellectual development. Professor Paul Arthur Schilpp, at that time at Northwestern University, in 1947–1948 invited a number of scholars to write essays on Albert Einstein: Philosopher-Scientist, *for which he persuaded Einstein to write his own intellectual autobiobraphy. This precious document, translated by Schilpp, gives us a unique view into the mind of Einstein at the end of his career.*

Here I sit in order to write, at the age of 67, something like my own obituary. . . .

Even when I was a fairly precocious young man the nothingness of the hopes and strivings which chases most men restlessly through life came to my consciousness with considerable vitality. Moreover, I soon discovered the cruelty of that chase, which in those years was much more carefully covered up by hypocrisy and glittering words than is the case today. By the mere existence of his stomach everyone was condemned to participate in that chase. Moreover, it was possible to satisfy the stomach by such participation, but not man in so far as he is a thinking and feeling being. As the first way out there was religion, which is implanted into every child by way of the traditional education-machine. Thus I came—despite the fact that I was the son of entirely irreligious (Jewish) parents—to a deep religiosity, which, however, found an abrupt ending at the age of 12. Through the reading of popular scientific books I soon reached the conviction that much in the stories of the Bible could not be true. The consequence was a positively fanatic [orgy of] freethinking coupled

SOURCE. Paul A. Schilpp, ed., *Albert Einstein: Philosopher-Scientist,* New York: Harper Torchbooks, 1959, Vol. I, pp. 3, 5, 7, 9, 11, 13, 19, 21, 25, 27, 29, 33, 35, 37, 51, 53, 55, and 57.

with the impression that youth is intentionally being deceived by the state through lies; it was a crushing impression. Suspicion against every kind of authority grew out of this experience, a skeptical attitude towards the convictions which were alive in any specific social environment—an attitude which has never again left me, even though later on, because of a better insight into the causal connections, it lost some of its original poignancy.

It is quite clear to me that the religious paradise of youth, which was thus lost, was a first attempt to free myself from the chains of the "merely personal," from an existence which is dominated by wishes, hopes and primitive feelings. Out yonder there was this huge world, which exists independently of us human beings and which stands before us like a great, eternal riddle, at least partially accessible to our inspection and thinking. The contemplation of this world beckoned like a liberation, and I soon noticed that many a man whom I had learned to esteem and to admire had found inner freedom and security in devoted occupation with it. The mental grasp of this extra personal world within the frame of the given possibilities swam as highest aim half consciously and half unconsciously before my mind's eye. Similarly motivated men of the present and of the past, as well as the insights which they had achieved, were the friends which could not be lost. The road to this paradise was not as comfortable and alluring as the road to the religious paradise; but it has proved itself as trustworthy, and I have never regretted having chosen it. . . .

What, precisely, is "thinking"? When, at the reception of sense-impressions, memory-pictures emerge, this is not yet "thinking." And when such pictures form series, each member of which calls forth another, this too is not yet "thinking." When, however, a certain picture turns up in many such series, then—precisely through such return—it becomes an ordering element for such series, in that it connects series which in themselves are unconnected. Such an element becomes an instrument, a concept. I think that the transition from free association or "dreaming" to thinking is characterized by the more or less dominating rôle which the "concept" plays in it. It is by no means necessary that a concept must be connected with a sensorily cognizable and reproducible sign (word); but when this is the case thinking becomes by means of that fact communicable.

With what right—the reader will ask—does this man operate so carelessly and primitively with ideas in such a problematic realm without making even the least effort to prove anything? My defense: all our thinking is of this nature of a free play with concepts; the justification for this play lies in the measure of survey over the experience of the senses which we are able to achieve with its aid. The concept of "truth" can not yet be applied to such a structure; to my thinking this concept can come in question only when a far-reaching agreement (*convention*) concerning the elements and rules of the game is already at hand.

For me it is not dubious that our thinking goes on for the most part without use of signs (words) and beyond that to a considerable degree unconsciously. For how, otherwise, should it happen that sometimes we "wonder" quite spontaneously about some experience? This "wondering" seems to occur when an experience comes into conflict with a world of concepts which is already sufficiently fixed in us. Whenever such a conflict is experienced hard and intensively it reacts back upon our thought world in a decisive way. The development of this thought world is in a certain sense a continuous flight from "wonder."

A wonder of such nature I experienced as a child of 4 or 5 years, when my father showed me a compass. That this needle behaved in such a determined way did not at all fit into the nature of events, which could find a place in the unconscious world of concepts (effect connected with direct "touch"). I can still remember—or at least believe I can remember—that this experience made a deep and lasting impression upon me. Something deeply hidden had to be behind things. What man sees before him from infancy causes no reaction of this kind; he is not surprised over the falling of bodies, concerning wind and rain, nor concerning the moon or about the fact that the moon does not fall down, nor concerning the differences between living and non-living matter.

At the age of 12 I experienced a second wonder of a totally different nature: in a little book dealing with Euclidean plane geometry, which came into my hands at the beginning of a school-year. Here were assertions, as for example the intersection of the three altitudes of a triangle in one point, which—though by no means evident—could nevertheless be proved with such certainty that any doubt appeared to be out of the question. This lucidity and certainty made an indescribable impression upon me. That the axiom had to be accepted unproved did not disturb me. In any case it was quite sufficient for me if I could peg proofs upon propositions the validity of which did not seem to me to be dubious. . . .

Now that I have allowed myself to be carried away sufficiently to interrupt my scantily begun obituary, I shall not hesitate to state here in a few sentences my epistemological credo, although in what precedes something has already incidentally been said about this. This credo actually evolved only much later and very slowly and does not correspond with the point of view I held in younger years.

I see on the one side the totality of sense-experiences, and, on the other, the totality of the concepts and propositions which are laid down in books. The relations between the concepts and propositions among themselves and each other are of a logical nature, and the business of logical thinking is strictly limited to the achievement of the connection between concepts and propositions among each other according to firmly laid down rules, which are the concern of logic. The concepts and propositions get "meaning," viz., "content," only through their connection with sense-experiences. The connection of the latter with the former is purely intuitive, not itself of a logical nature. The degree of certainty with which this connection, viz., intuitive combination, can be undertaken, and nothing else, differentiates empty phantasy from scientific "truth." The system of concepts is a creation of man together with the rules of syntax, which constitute the structure of the conceptual systems. Although the conceptual systems are logically entirely arbitrary, they are bound by the aim to permit the most nearly possible certain (intuitive) and complete co-ordination with the totality of sense-experiences; secondly they aim at greatest possible sparsity of their logically independent elements (basic concepts and axioms), i.e., undefined concepts and underived [postulated] propositions.

A proposition is correct if, within a logical system, it is deduced according to the accepted logical rules. A system has truth-content according to the certainty and completeness of its co-ordination-possibility to the totality of experience. A correct proposition borrows its "truth" from the truth-content of the system to which it belongs. . . .

Now to the field of physics as it presented itself at that time [1896]. In spite of all the fruitfulness in particulars, dogmatic rigidity prevailed in matters of principles: In the beginning (if there was such a thing) God created Newton's laws of motion together with the necessary masses and forces. This is all; everything beyond this follows from the development of appropriate mathematical methods by means of deduction. What the nineteenth century achieved on the strength of this basis, especially through the application of the partial differential equations, was bound to arouse the admiration of every receptive person. . . .

We must not be surprised, therefore, that, so to speak, all physicists of the last century saw in classical mechanics a firm and final foundation for all physics, yes, indeed, for all natural science, and that they never grew tired in their attempts to base Maxwell's theory of electro-magnetism, which, in the meantime, was slowly beginning to win out, upon mechanics as well. Even Maxwell and H. Hertz, who in retrospect appear as those who demolished the faith in mechanics as the final basis of all physical thinking, in their conscious thinking adhered throughout to mechanics as the secured basis of physics. It was Ernst Mach who, in his *History of Mechanics*, shook this dogmatic faith; this book exercised a profound influence upon me in this regard while I was a student. I see Mach's greatness in his incorruptible skepticism and independence; in my younger years, however, Mach's epistemological position also influenced me very greatly, a position which today appears to me to be essentially untenable. For he did not place in the correct light the essentially constructive and speculative nature of thought and more especially of scientific thought; in consequence of which he condemned theory on precisely those points where its constructive-speculative character unconcealably comes to light, as for example in the kinetic atomic theory. . . .

And now to the critique of mechanics as the basis of physics.

From the first point of view (confirmation by experiment) the incorporation of wave-optics into the mechanical picture of the world was bound to arouse serious misgivings. If light was to be interpreted as undulatory motion in an elastic body (ether), this had to be a medium which permeates everything; because of the transversality of the lightwaves in the main similar to a solid body, yet incompressible, so that longitudinal waves did not exist. This ether had to lead a ghostly existence alongside the rest of matter, inasmuch as it seemed to offer no resistance whatever to the motion of "ponderable" bodies. In order to explain the refraction-indices of transparent bodies as well as the processes of emission and absorption of radiation, one would have had to assume complicated reciprocal actions between the two types of matter, something which was not even seriously tried, let alone achieved.

Furthermore, the electromagnetic forces necessitated the introduction of electric masses, which, although they had no noticeable inertia, yet interacted with each other, and whose interaction was, moreover, in contrast to the force of gravitation, of a polar type.

The factor which finally succeeded, after long hesitation, to bring the physicists slowly around to give up the faith in the possibility that all of physics could be founded upon Newton's mechanics, was the electrodynamics of Faraday and Maxwell. For this theory and its confirmation by Hertz's experiments showed that there are electromagnetic phenomena which by their very nature are detached from every ponderable matter—namely the waves in empty space which consist of electromagnetic "fields." If mechanics was to be maintained as the foundation of physics, Maxwell's equations had to be interpreted mechanically. This was zealously but fruitlessly attempted, while the equations were proving themselves fruitful in mounting degree. One got used to operating with these fields as independent substances without finding it necessary to give one's self an account of their mechanical nature; thus mechanics as the basis of physics was being abandoned, almost unnoticeably, because its adaptability to the facts presented itself finally as hopeless. Since then there exist two types of conceptual elements, on the one hand, material points with forces at a distance between them, and, on the other hand, the continuous field. It presents an intermediate state in physics without a uniform basis for the entirety, which—although unsatisfactory—is far from having been superseded. . . .

Now for a few remarks to the critique of mechanics as the

foundation of physics from the second, the "interior," point of view. In today's state of science, i.e., after the departure from the mechanical foundation, such critique has only an interest in method left. But such a critique is well suited to show the type of argumentation which, in the choice of theories in the future will have to play an all the greater rôle the more the basic concepts and axioms distance themselves from what is directly observable, so that the confrontation of the implications of theory by the facts becomes constantly more difficult and more drawn out. First in line to be mentioned is Mach's argument, which, however, had already been clearly recognized by Newton (bucket experiment). From the standpoint of purely geometrical description all "rigid" co-ordinate systems are among themselves logically equivalent. The equations of mechanics (for example this is already true of the law of inertia) claim validity only when referred to a specific class of such systems, i.e., the "inertial systems." In this the co-ordinate system as bodily object is without any significance. It is necessary, therefore, in order to justify the necessity of the specific choice, to look for something which lies outside of the objects (masses, distances) with which the theory is concerned. For this reason "absolute space" as originally determinative was quite explicitly introduced by Newton as the omnipresent active participant in all mechanical events; by "absolute" he obviously means uninfluenced by the masses and by their motion. What makes this state of affairs appear particularly offensive is the fact that there are supposed to be infinitely many inertial systems, relative to each other in uniform translation, which are supposed to be distinguished among all other rigid systems.

Mach conjectures that in a truly rational theory inertia would have to depend upon the interaction of the masses, precisely as was true for Newton's other forces, a conception which for a long time I considered as in principle the correct one. It presupposes implicitly, however, that the basic theory should be of the general type of Newton's mechanics: masses and their interaction as the original concepts. The attempt at such a solution does not fit into a consistent field theory, as will be immediately recognized. . . .

Reflections of this type made it clear to me as long ago as

shortly after 1900, . . . that neither mechanics nor thermodynamics could (except in limiting cases) claim exact validity. By and by I despaired of the possibility of discovering the true laws by means of constructive efforts based on known facts. The longer and the more despairingly I tried, the more I came to the conviction that only the discovery of a universal formal principle could lead us to assured results. The example I saw before me was thermodynamics. The general principle was there given in the theorem: the laws of nature are such that it is impossible to construct a *perpetuum mobile* (of the first and second kind). How, then, could such a universal principle be found? After ten years of reflection such a principle resulted from a paradox upon which I had already hit at the age of sixteen: If I pursue a beam of light with the velocity c (velocity of light in a vacuum), I should observe such a beam of light as a spatially oscillatory electromagnetic field at rest. However, there seems to be no such thing, whether on the basis of experience or according to Maxwell's equations. From the very beginning it appeared to me intuitively clear that, judged from the standpoint of such an observer, everything would have to happen according to the same laws as for an observer who, relative to the earth, was at rest. For how, otherwise, should the first observer know, i.e., be able to determine, that he is in a state of fast uniform motion?

One sees that in this paradox the germ of the special relativity theory is already contained. Today everyone knows, of course, that all attempts to clarify this paradox satisfactorily were condemned to failure as long as the axiom of the absolute character of time, viz., of simultaneity, unrecognizedly was anchored in the unconscious. Clearly to recognize this axiom and its arbitrary character really implies already the solution of the problem. The type of critical reasoning which was required for the discovery of this central point was decisively furthered, in my case, especially by the reading of David Hume's and Ernst Mach's philosophical writings.

One had to understand clearly what the spatial co-ordinates and the temporal duration of events meant in physics. The physical interpretation of the spatial co-ordinates presupposed a fixed body of reference, which, moreover, had to be in a more or less definite state of motion (inertial system). In a given inertial system the co-ordinates meant the results of certain measurements with rigid (stationary) rods. (One should always be conscious of the fact that the presupposition of the existence in principle of rigid rods is a presupposition suggested by approximate experience, but which is, in principle, arbitrary.) With such an interpretation of the spatial co-ordinates the question of the validity of Euclidean geometry becomes a problem of physics.

If, then, one tries to interpret the time of an event analogously, one needs a means for the measurement of the difference in time (in itself determined periodic process realized by a system of sufficiently small spatial extension). A clock at rest relative to the system of inertia defines a local time. The local times of all space points taken together are the "time," which belongs to the selected system of inertia, if a means is given to "set" these clocks relative to each other. One sees that *a priori* it is not at all necessary that the "times" thus defined in different inertial systems agree with one another. One would have noticed this long ago, if, for the practical experience of everyday life light did not appear (because of the high value of c), as the means for the statement of absolute simultaneity.

The presupposition of the existence (in principle) of (ideal, viz., perfect) measuring rods and clocks is not independent of each other; since a lightsignal, which is reflected back and forth between the ends of a rigid rod, constitutes an ideal clock, provided that the postulate of the constancy of the light-velocity in vacuum does not lead to contradictions.

The above paradox may then be formulated as follows. According to the rules of connection, used in classical physics, of the spatial co-ordinates and of the time of events in the transition from one inertial system to another the two assumptions of (1) the constancy of the light velocity, and (2) the independence of the laws (thus specially also of the law of the constancy of the light velocity) of the choice of the inertial system (principle of special relativity) are mutually incompatible (despite the fact that both taken separately are based on experience).

The insight which is fundamental for the special theory of relativity is this: The assumptions (1) and (2) are compatible if relations of a new type ("Lorentz-transformation") are postulated for the conversion of co-ordinates and the times of events.

With the given physical interpretation of co-ordinates and time, this is by no means merely a conventional step, but implies certain hypotheses concerning the actual behavior of moving measuring-rods and clocks, which can be experimentally validated or disproved.

The universal principle of the special theory of relativity is contained in the postluate: The laws of physics are invariant with respect to the Lorentz-transformations (for the transition from one inertial system to any other arbitrarily chosen system of inertia). This is a restricting principle for natural laws, comparable to the restricting principle of the non-existence of the *perpetuum mobile* which underlies thermodynamics.

2 FROM *Sir Edmund Whittaker* *A History of the Theories of Aether and and Electricity*

Sir Edmund Whittaker, Fellow of the Royal Society, turned to the writing of the history of science after a distinguished career as a theoretical physicist. His views of the origins of Relativity Theory are those of a man who lived through the revolution in thought that he describes.

THE RELATIVITY THEORY OF POINCARÉ AND LORENTZ

At the end of the nineteenth century, one of the most perplexing unsolved problems of natural philosophy was that of determining the relative motion of the earth and the aether. Let

SOURCE. Sir Edmund Whittaker, *A History of the Theories of Aether and Electricity*, Vol. II: *The Modern Theories*, London: Thomas Nelson and Sons Ltd., 1953, pp. 27–31, 36, 40, and 42–43. Reprinted by permission of the publisher.

us try to present the matter as it appeared to the physicists of that time. . . .

In the latter part of the nineteenth century the doctrine of the aether, which was justified by the undulatory theory of light, was generally regarded as involving the concepts of rest and motion relative to the aether, and thus to afford a means of specifying absolute position and defining the Body Alpha. Suppose, for instance, that a disturbance is generated at any point in free aether : this disturbance will spread outwards in the form of a sphere : and the centre of this sphere will for all subsequent time occupy an unchanged position relative to the aether. In this way, or in many other ways, we might hope to determine, by electrical or optical experiments, the velocity of the earth's motion relative to the aether.

In the first years of the twentieth century this problem was provoking a fresh series of experimental investigations. The most interesting of these was due to FitzGerald who, shortly before his death in February 1901, commenced to examine the phenomena exhibited by a charged electrical condenser, as it is carried through space by the terrestrial motion. When the plane of the condenser includes the direction of the aether-drift (the "longitudinal position"), the moving positive and negative charges on its two plates will be equivalent to currents running tangentially in opposite directions in the plates, so that a magnetic field will be set up in the space between them, and magnetic energy must be stored in this space : but when the plane of the condenser is at right angles to the terrestrial motion (the "transverse position"), the equivalent currents are in the normal direction, and neutralise each other's magnetic action almost completely. FitzGerald's original idea was that, in order to supply the magnetic energy, there must be a mechanical drag on the condenser at the moment of charging, similar to that which would be produced if the mass of a body at the surface of the earth were suddenly to become greater. Moreover, the co-existence of the electric and magnetic fields in the space between the plates would entail the existence of an electromagnetic momentum proportional to their vector-product. This momentum is easily seen to be (with sufficient approximation) parallel to the plates, and so would not in general have the same direction as the velocity of the condenser relative to the aether: thus the change in the situation in one second might be represented by the annihilation of the momentum existing at the beginning of the second and the creation of the momentum (equal and parallel to it) existing at the end of the second. But two equal and oppositely-parallel momenta at a distance apart constitute an angular momentum: and we may therefore expect that if the condenser is freely suspended, there will in general be a couple acting on it, proportional to the vector-product of the velocity of the condenser and the electromagnetic momentum. This couple would vanish in either the longitudinal or the tranverse orientation, but in intermediate positions would tend to rotate the condenser into the longitudinal position; the tranverse position would be one of unstable equilibrium.

For both effects a search was made by FitzGerald's pupil F. T. Trouton; in the experiments designed to observe the turning couple, a condenser was suspended in a vertical plane by a fine wire, and charged. The effect to be detected was small: for the magnetic force due to the motion of the charges would be of order (w/c), where w denotes the velocity of the earth: so the magnetic energy of the system, which depends on the square of the force, would be of order $(w/c)^2$: and the couple would likewise be of the second order in (w/c).

No effect of any kind could be detected, a result whose explanation was rightly surmised by P. Langevin to belong to the same order of ideas as FitzGerald's hypothesis of contraction.

It may be remarked that the existence of the couple, had it been observed, would have demonstrated the possibility of drawing on the energy of the earth's motion for purposes of terrestrial utility.

The FitzGerald contraction of matter as it moves through the aether might conceivably be supposed to affect in some way the optical properties of the moving matter: for instance, transparent substances might become doubly refracting. Experiments designed to test this supposition were performed by Lord Rayleigh in 1902 and by D. B. Brace in 1904, but no double refraction comparable with the proportion $(w/c)^2$ of the single refraction could be detected. The FitzGerald contraction of a material body cannot therefore be of the same nature as the contraction which would be produced in the body by pressure, but must be accompanied by such concomitant changes in the relations of the molecules to the aether, that an isotropic substance does not lose its simply refracting character.

Even before the end of the nineteenth century, the failure of so many promising attempts to measure the velocity of the earth relative to the aether had suggested to the penetrating and original mind of Poincaré a new possibility. In his lectures at the Sorbonne in 1899, after describing the experiments so far made, which had yielded no effects involving either the first or the second powers of the coefficient of aberration (i.e. the ratio of the earth's velocity to the velocity of light), he went on to say, "I regard it as very probable that optical phenomena depend only on the *relative* motions of the material bodies, luminous sources, and optical apparatus concerned, and that this is true not merely as far as quantities of the order of the square of the aberration, but *rigorously*." In other words, Poincaré believed in 1899 that *absolute motion is indetectible in principle*, whether by dynamical, optical, or electrical means.

In the following year, at an International Congress of Physics held at Paris, he asserted the same doctrine. "Our aether," he said, "does it really exist? I do not believe that more precise observations could ever reveal anything more than *relative* displacements." After referring to the circumstance that the explanations then current for the negative results regarding terms of the first order in (w/c) were different from the explanations regarding the second order terms, he went on, "It is necessary to find the *same* explanation for the negative results obtained regarding terms of these two orders: and there is every reason to suppose that this explanation will then apply equally to terms of higher orders, and that the mutual destruction of the terms will be rigorous and absolute." A new principle would thus be introduced into physics, which would resemble the Second Law of Thermodynamics in as much as it asserted the *impossibility of doing something*: in this case, the impossibility of determining the velocity of the earth relative to the aether.

In a lecture to a Congress of Arts and Science at St Louis, U.S.A., on 24 September 1904, Poincaré gave to a generalised form of this principle the name, *The Principle of Relativity*.

"According to the Principle of Relativity," he said, "the laws of physical phenomena must be the same for a 'fixed' observer as for an observer who has a uniform motion of translation relative to him : so that we have not, and cannot possibly have, any means of discerning whether we are, or are not, carried along in such a motion." After examining the records of observation in the light of this principle, he declared, "From all these results there must arise an entirely new kind of dynamics, *which will be characterised above all by the rule, that no velocity can exceed the velocity of light*". . . .

Some of the consequences of the new theory seemed to contemporary physicists very strange. Suppose, for example, that two inertial sets of axes A and B are in motion relative to each other, and that at a certain instant their origins coincide : and suppose that at this instant a flash of light is generated at the common origin. Then, by what has been said in the subsequent propagation, the wave-fronts of the light, as observed in A and in B, are spheres whose centres are the origins of A and B respectively, and therefore *different* spheres. How can this be?

The paradox is explained when it is remembered that a wave-front is defined to be the locus of points which are *simultaneously* in the same phase of disturbance. Now events taking place at different points, which are simultaneous according to A's system of measuring time, are not in general simultaneous according to B's way of measuring : and therefore what A calls a wave-front is not the same thing as what B calls a wave-front. Moreover, since the system of measuring space is different in the two inertial systems, what A calls a sphere is not the same thing as what B calls a sphere. Thus there is no contradiction in the statement that the wave-fronts for A are spheres with A's origin as centre, while the wave-fronts for B are spheres with B's origin as centre.

In common language we speak of events which happen at different points of space as happening "at the same instant of time," and we also speak of events which happen at different instants of time as happening "at the same point of space." We now see that such expressions can have a meaning only by virtue of artificial conventions; they do not correspond to any essential physical realties.

It is usual to regard Poincaré as primarily a mathematician, and Lorentz as primarily a theoretical physicist: but as regards their contributions to relativity theory, the positions were reversed: it was Poincaré who proposed the general physical principle, and Lorentz who supplied much of the mathematical embodiment. Indeed, Lorentz was for many years doubtful about the physical theory: in a lecture which he gave in October 1910 he spoke of "die Vorstellung (die auch Redner nur ungern aufgeben würde), dass Raum und Zeit etwas vollig Verschiedenes seien und dass es eine 'wahre Zeit' gebe (die Gleichzeitigkeit würde denn unabhängig vom Orte bestehen)."*

A distinguished physicist who visited Lorentz in Holland shortly before his death found that his opinions on this question were unchanged. . . .

In the autumn of 1905, in the same volume of the *Annalen der Physik* as his paper on the Brownian motion, Einstein published a paper which set forth the relativity theory of Poincaré and Lorentz with some amplifications, and which attracted much attention. He asserted as a fundamental principle the *constancy of the velocity of light*, i.e. that the velocity of light *in vacuo* is the same in all systems of reference which are moving relatively to each other: an assertion which at the time was widely accepted, but has been severely criticised by later writers. . . .

It is clear, from the history set forth in the present chapter, that the theory of relativity had its origin in the theory of aether and electrons. When relativity had become recognised as a doctrine covering the whole operation of physical nature, efforts were made to present it in a form free from any special association with electromagnetic theory, and deducible logically from a definite set of axioms of greater or less plausibility.

It should be mentioned also that when relativity theory had become generally accepted, the Michelson-Morley experiment was rediscussed with a much more complete understanding and exactitude.

* "The concept (which the present author would dislike to abandon) that space and time are something completely distinct and that a 'true time' exists (simultaneity would then have a meaning independent of position)."

3 *Gerald Holton*
"On the Origins of the Special Theory of Relativity"

Gerald Holton, Professor of Physics at Harvard University, is preparing a major study on Albert Einstein. He looks at the historical background to Relativity Theory in a somewhat different way from Whittaker. Here Holton, as a historian of science, points out the many dangers the scientist, if unschooled in historical method, faces when he sets out to write the history of a major scientific event.

When I received the persuasive invitation to speak today on a problem of theory construction and of the logic of discovery, I noted particularly the request to bring out the historical-sociological aspects. This directive was a pleasant surprise, for I recalled that Hans Reichenbach had flatly declared himself for the opposite view when he said "The philosopher of science is not much interested in the thought processes which lead to scientific discoveries . . . , that is, he is not interested in the context of discovery, but in the context of justification." If, therefore, I shall make some remarks on the origins of Einstein's special theory of relativity, I will be disobeying the Reichenbachian dictum. However, I draw further strength for this resolution from Einstein, who himself declared for the value of the historical treatment of the rise of key theories in science. . . .

I speak of Einstein's work because his case is both typical and special. The rise of relativity theory shares many features with the rise of other important scientific theories in our time, and in addition it is of course very much more: To find another work that illuminates as richly the relationship between physics, mathematics, and epistemology, or between experiment and theory, or one with the same range of scientific, philosophical and gen-

SOURCE. Gerald Holton, "On the Origins of the Special Theory of Relativity," *American Journal of Physics*, Vol. 28, Nos. 1–9, 1960, pp. 627–631 and 633–636 (not including footnotes). This is a condensation of a longer paper. Reprinted by permission of the publisher and the author.

eral intellectual implications, one would have to go back to Newton's *Principia.* The theory of relativity was a key development, both in physical science itself and also in modern philosophy of science. The reason for its dual significances is that Einstein's work provided not only a new principle of physics, but, as A. N. Whitehead said, "a principle, a procedure, and an explanation." Accordingly, the commentaries on the historical origins of the theory of relativity have tended to fall into two classes, each having distinguished proponents: the one views it as a mutant, a sharp break with respect to the work of the immediate predecessors of Einstein; the other regards it as an elaboration of then current work, e.g., by Lorentz and Poincaré.

To my mind, the Einsteinian innovation is understood best by superposition of both views, by seeing the discontinuity of methodological orientation within an historically continuous scientific development. But before we come to discuss this, and if we take seriously my point of view, we should first be ready to investigate a number of real problems of the historical or even "historical-sociological" kind: What are the sources for a study of the origins of the special theory of relativity (*RT*) and what is their probable reliability? What was the state of science around 1905, what were the contributions which prepared the field for the *RT*, and what did Einstein know about them? What were the steps by which Einstein reached the conclusions he published in 1905? To what extent was this work a member of a continuous chain having as its immediate predecessors Lorentz and Poincaré? What was the role of experiment in the genesis of the *RT*, and what the role of the existence of contradictory hypotheses? What part played epistemological analysis in Einstein's thought? What was the early reception of the *RT* among scientists? In particular, what was Einstein's relation with Mach, Lorentz, and Planck? What may we say about the style of Einstein's work and his personal orientations? What, if anything, in the origins and content of the RT is typical of other theories with great impact on science? And even, what methodological principles for the study of the history of science emerge from this study? . . .

When one studies the relativity papers in the larger contextual setting of Einstein's other scientific papers, particularly those on the quantum theory of light and on Brownian motion which also were written and published in 1905, one notices two crucial points. While the three epochal papers of 1905—sent to the *Annalen der Physik* at intervals of less than eight weeks—seem to be in entirely different fields, closer study shows that they arose in fact from the same general problem, namely, the fluctuations in the pressure of radiation. In 1905, as Einstein later wrote to von Laue, he had already known that Maxwell's theory leads to the wrong prediction of the motion of a delicately suspended mirror "in a Planckian radiation cavity." This connects on the one hand with the consideration of Brownian motion as well as to the quantum structure of radiation, and on the other hand with Einstein's more general reconsideration of "the electromagnetic foundations of physics" itself.

One also finds that the style of the three papers is essentially the same, and reveals what is typical of Einstein's work at that time. Each begins with the statement of formal asymmetries or other incongruities of a predominantly esthetic nature (rather than, for example, a puzzle posed by unexplained experimental facts), then proposes a principle—preferably one of the generality of, say, the second law of thermodynamics, to cite Einstein's repeated analogy—which removes the asymmetries as one of the deduced consequences, and at the end produces one or more experimentally verifiable predictions. . . .

The third paper of 1905 is, of course, Einstein's first paper on the *RT*. He begins again by drawing attention to a formal asymmetry, i.e., in the description of currents generated during relative motion between magnets and conductors. The paper does not invoke explicitly any of the several well-known experimental difficulties—and the Michelson and Michelson-Morley experiments are not even mentioned when the opportunity arises to show in what manner the *RT* accounts for them. . . .

RETURN TO A CLASSIC RESTRICTION ON HYPOTHESES

The recognition of these common elements in the three papers prepares us for the essential realization that the fundamental postulates appearing in each of the three papers are *heuristic*. The heuristic nature of the postulate of relativity was from the beginning apparent to Einstein (as he asserted in 1907 and later) because of the restriction of the *RT* to translational motions and to gravitation-free space.

The study of the three papers together reveals also the extent to which Einstein's *RT* represents an attempt to restrict hypotheses to the most *general kind* and the *smallest number* possible —a goal on which Einstein often insisted. In the 1905 paper on *RT*, he makes, in addition to the two "conjectures" raised to "postulates" (i.e., of relativity and of the constancy of light velocity) only four other hypotheses: one of the isotropy and homogeneity of space, the others concerning three logical properties of the definition of synchronization of watches. In contrast, H. A. Lorentz's great paper which appeared a year before Einstein's publication and typified the best work in physics of its time—a paper which Lorentz declared to be based on "fundamental assumptions" rather than on "special hypotheses"—contained in fact eleven *ad hoc* hypotheses: restriction to small ratios of velocities v to light velocity c; postulation *a priori* of the transformation equations (rather than their derivation from other postulates); assumption of a stationary ether; assumption that the stationary electron is round; that its charge is uniformly distributed; that all mass is electromagnetic; that the moving electron changes one of its dimensions precisely in the ratio of $(1 - v^2/c^2)^{1/2}$ to 1; that forces between uncharged particles and between a charged and uncharged particle have the same transformation properties as electrostatic forces in the electrostatic system; that all charges in atoms are in a certain number of separate "electrons"; that each of these is acted on only by others in the same atom; and that atoms in motion as a whole deform as electrons themselves do. It is for these reasons that Einstein later maintained that the *RT* grew out of the Maxwell-Lorentz theory of electrodynamics "as an amazingly simple summary and generalization of hypotheses which previously have been independent of one another. . . ."

If one has studied the development of scientific theories, one notes here a familiar theme: *the so-called scientific "revolution" turns out to be at bottom an effort to return to a classical purity.*

This is not only a key to a new evaluation of Einstein's contribution, but indicates a fairly general characteristic of great scientific "revolutions." Indeed, while it is usually stressed that Einstein challenged Newtonian physics in fundamental ways, the equally correct but neglected point is the number of methodological correspondences with earlier classics, for example, with the *Principia*. . . .

WHITTAKER'S ACCOUNT OF THE ORIGINS OF EINSTEIN'S WORK

. . . I wish to turn to a question on which a dispute has been active, namely, to what extent Einstein's work was original rather than anticipated by, or specifically based on, other published work. Particularly interesting is the essay on Einstein by Sir Edmund Whittaker in the *Biographical Memoirs of Fellows of the Royal Society* (London, 1955). Whittaker's commitment to the 19th-century tradition of physics and to the ether theory is illustrated in his well-known book *A History of the Theories of Aether and Electricity* up to about 1900 (London, 1910; 2nd ed., 1951), and also by his excellent contributions in the field of classical mechanics. Moreover, in the second volume of the *History*, completed in 1953, which carries the story to 1926, Whittaker had largely dismissed Einstein's paper of 1905 on the *RT* as one "which set forth the relativity theory of Poincaré and Lorentz with some amplifications, and which attracted much attention."

This presentation evoked considerable criticism, some of which I know to have reached Whittaker while his book was still in manuscript, and some of which reached him by the time he composed the biographical memoir after Einstein's death in 1955. It is therefore noteworthy that in his 1955 necrolog for Einstein, Whittaker has not changed his earlier evaluation. For example, he repeats that Poincaré in a speech in St. Louis, U.S.A., in September 1904 had coined the phrase "principle of relativity." Whittaker asks how physics could have been reformulated in accordance with "Poincaré's principle of relativity," and he reports that with respect to the laws of the electromagnetic field this "discovery was made in 1903 by Lorentz," citing a paper by Lorentz in the *Proceedings of the Academy of Sciences, Amsterdam,* for the year 1903. Whittaker shows that "the fundamental equations of the aether in empty space" are invariant under suitably chosen (i.e., Lorentz) transformations, and he concludes with the remarkable sentence: "Einstein in [the *RT* paper of 1905] adopted Poincaré's principle of relativity, using Poincaré's name for it, as a new basis for physics and showed that the group of Lorentz transformations provided a new analysis connecting the physics of bodies in motion relative to each other."

Since Whittaker's analysis has been and is likely to continue to be given considerable weight, it is necessary to examine it closely. It turns out to be an excellent example of the proposition that no such analysis can be considered meaningful except insofar as it deals both with the material it purports to cover *and* with the prior commitments and prejudices of the scholar himself. Here is a brief summary of main findings when Whittaker's analysis is considered in this light.

(1) Einstein's *RT* paper of 1905 was indeed one of a number of contributions by many different authors in the general field of the electrodynamics of moving bodies. In the *Annalen der Physik* alone there are eight papers from 1902 up to 1905 concerned with this general problem. Einstein himself always insisted on this aspect of continuity. The earliest evidence is a letter written in the spring of 1905 to his friend Conrad Habicht, describing his various investigations. In one sentence he describes the developing *RT* paper: "The fourth work lies at hand in draft [*liegt im Konzept vor*] and is an electrodynamics of moving bodies making use of a *modification* of the theory of space and time; you will surely be interested in the purely kinematic part of this work." Seelig also quotes a later remark of Einstein which gives in one sentence his often repeated attitude: "With respect to the theory of relativity it is not at all a question of a revolutionary act, but of a natural development of a line which can be pursued through centuries"....

(2) The paper by Poincaré of 1904 which Whittaker cites turns out not to enunciate the new relativity principle, but is rather a very acute and penetrating though qualitative summary of the difficulties which contemporary physics was then making for six classical laws or principles, including what is in effect the

Galilean-Newtonian principle of relativity. . . . Of the principle of relativity Poincaré complains that it "is battered" by current developments in electromagnetic theory, although, he says, it "is confirmed by daily experience" and "is imposed in an irresistible way upon one's good sense." Poincaré's main point is to show the need for a new development, the outlines of which he suggests in these words: "Perhaps likewise we should construct a whole new mechanics, that we only succeed in catching a glimpse of, where inertia increasing with the velocity, the velocity of light would become an impassable limit." Thus he illustrates both the power of his intuition and the qualitative nature of the suggestion.

(3) It is more difficult to discuss the 1903 paper of Lorentz which Whittaker, both in his book and in his Memoir, cited specifically as the work that spelled out most of the basic details of Einstein's *RT* of 1905. In the first place, this paper does not exist. What Whittaker clearly wished to refer to is the paper Lorentz published a year later, in 1904. Since Whittaker was otherwise very careful with the voluminous citations of references, this repeated slip, which doubles the time interval between the work of Lorentz and of Einstein, is not merely a mistake. It is at least a symbolic mistake–symbolic of the way a biographer's preconceptions interact with his material.

(4) Whittaker clearly implied that Einstein used Lorentz's transformation equation published in 1904. He therefore chose to neglect that both Einstein and those close to him have repeatedly said that Einstein had not read Lorentz's 1904 paper.

(5) Even if one does not wish to rely on the word of Einstein and other prominent physicists of his time in this matter, there are four items of internal evidence in Einstein's 1905 paper which indicate that he had not read Lorentz's of 1904. Einstein does write the transformation equations in a form equivalent of those of Lorentz (or, for that matter, of Voigt's of 1887); but whereas Lorentz had assumed these equations *a priori* in order to obtain the invariance of Maxwell's equations in free space, Einstein *derived* them from the two fundamental postulates of the *RT*. He therefore did not need to know of Lorentz's paper of 1904. . . .

(6) Quite apart from the question whether Einstein's 1905 paper was written independently of Lorentz's is the equally sig-

nificant fact that in a crucial sense Lorentz's paper was of course not on the relativity theory as we understand the term since Einstein. Lorentz's fundamental assumptions are not relativistic; as Born says, "he never claimed to be the author of the principle of relativity," and, on the contrary, referred to it as "Einstein's Relativitätsprinzip" in his lectures of 1910. In Lorentz's essay on "The Principle of Relativity of Uniform Translation," published in 1922, six years before Lorentz's death, he still asked that space be considered to have "a certain substantiality; and if so, one may, in all modesty, call true time the time measured by clocks which are fixed in this medium, and consider simultaneity as a primary concept." In his 1904 paper he had postulated the nonrelativistic addition theorem for velocities, $v = V + u$, and even in the 1922 book he did not consider the velocity of light as inherently the highest attainable velocity of material bodies. . . .

In closing, I return to my initial remarks: The detailed study of the historical situation is, to my mind, an important first step in those discussions which try to base epistemological considerations on "real" cases. This is not always done easily; but it is through the dispassionate examination of historically valid cases that we can best become aware of the preconceptions which underlie all philosophical study.

4 *Adolf Grünbaum*
"The Genesis of the Special Theory of Relativity"

Adolf Grünbaum is Mellon Professor of the Philosophy of Science at the University of Pittsburgh. He views the origins of Relativity Theory from the vantage point of the philosopher and challenges both Whittaker's and Holton's account on philosophical grounds.

SOURCE. Herbert Feigl and Grover Maxwell, eds., *Current Issues in the Philosophy of Science*, New York: Holt, Rinehart and Winston, 1961, pp. 43–50.

The genesis of the special theory of relativity (hereafter denoted by "RT") invites investigation as part of any endeavor to view scientific theory construction in both logical and historical perspective. For in the case of RT there are two points whose consideration proves illuminating: (i) the relevance of knowledge of its *history* to the provision of an analysis of its *logical foundations*, and (ii) the capability of philosophical mastery of its logical foundations to preclude such *historical* errors as E. T. Whittaker's incorrect assessment of Einstein's role in the genesis of RT. . . .

If the history of the RT is to serve as a propaedeutic to the student of that theory's epistemological foundations, then the following key problem beckons detailed investigation by the historians: what considerations prompted Einstein to make the *two* physical assumptions which are at the root of his philosophical doctrine of the *conventionality* of the simultaneity of spatially separated events? This doctrine is set forth *very concisely* in the first paragraph of his 1905 paper, which is entitled "Definition of Simultaneity." There he writes, "We have not defined a common 'time' for [the spatially separated points] *A* and *B*, for the latter cannot be defined at all unless we establish *by definition* that the 'time' required by light to travel from *A* to *B* equals the 'time' it requires to travel from *B* to *A*."

The *two* physical assumptions on which this conception of simultaneity rests are the following:

(i) Within the class of physical events, material clocks do *not* define relations of absolute simultaneity under transport: if two clocks U_1 and U_2 are initially synchronized at the *same* place *A* and then transported via paths of *different lengths* to a different place *B* such that their arrivals at B coincide, then U_1 and U_2 will no longer be synchronized at *B*. And if U_1 and U_2 were brought to *B* via the *same* path (or via different paths which are of *equal* length) such that their arrivals do *not* coincide, then their initial synchronization would likewise be destroyed.

(ii) Light is the fastest signal *in vacuo* in the following topological sense: no kind of causal chain (moving particles, radiation) emitted *in vacuo* at a given point *A* together with a light pulse can reach any other point *B* earlier—as judged by a local clock at *B* which merely orders events there in a metrically arbitrary fashion—than this light pulse.

As we shall see presently, Einstein himself gives us tantalizingly incomplete explicit information concerning the grounds for his original confidence in his intuition that assumption (ii) is true. But even if we did possess full clarity on that score, we still confront the same question in regard to assumption (i) and are compelled to try to answer it with even less assistance from Einstein himself, as we shall see. The importance of understanding the grounds on which Einstein thought he could safely make assumption (i) can be gauged by the following basic fact: if assumption (i) had been thought to be *false*, then the belief in the truth of (ii) would *not* have warranted the abandonment of the received Newtonian doctrine of absolute simultaneity. And, in that eventuality, the members of the scientific community to whom Einstein addressed his paper of 1905 would have been fully entitled to *reject* his *conventionalist* conception of *one*-way transit times and velocities. But this conception is absolutely fundamental to his principle of the constancy of the speed of light, as is evident from the second paragraph of his 1905 paper, and it rests on his denial of absolute simultaneity. In fact, (i) is essential, because a physicist brought up in the Newtonian tradition quite naturally uses *not* signal connectibility but the readings of suitably transported clocks as the fundamental indicators of temporal order. He recognizes, of course, that the truth of (ii) compels such far-reaching revisions in his theoretical edifice as the repudiation of the second law of motion. But he stoutly and rightly maintains that *if* (i) is *false*, absolute simultaneity remains intact, unencumbered by the truth of (ii).

Specifically what does Einstein himself tell us about his original grounds for assuming the truth (ii)? At this point, it is essential to quote him *in extenso*. He writes:

"By and by I despaired of the possibility of discovering the true laws by means of constructive efforts based on known facts. The longer and the more despairingly I tried, the more I came to the conviction that only the discovery of a universal formal principle could lead us to assured results. The example I saw before me was thermodynamics. The general principle was there given in the theorem: the laws of nature are such that it is impossible to construct a *perpetuum mobile* (of the first and second kind). How, then, could such a universal principle be found? After ten years of reflection such a principle resulted from a paradox upon which I had already hit at the age of sixteen: If I pursue a beam of light with the velocity c (velocity of light in a vacuum), I should observe such a beam of light as a spatially oscillatory electromagnetic field at rest. However, there seems to be no such thing, whether on the basis of experience or according to Maxwell's equations. From the very beginning it appeared to me intuitively clear that, judged from the standpoint of such an observer, everything would have to happen according to the same laws as for an observer who, relative to the earth, was at rest. For how, otherwise, should the first observer know, that is, be able to determine, that he is in a state of fast uniform motion?

"One sees that in this paradox the germ of the special relativity theory is already contained. Today everyone knows, of course, that all attempts to clarify this paradox satisfactorily were condemned to failure as long as the axiom of the absolute character of time, viz., of simultaneity, unrecognizedly was anchored in the unconscious. Clearly to recognize this axiom and its arbitrary character really implies already the solution of the problem. The type of critical reasoning which was required for the discovery of this central point was decisively furthered, in my case, especially by the reading of David Hume's and Ernst Mach's philosophical writings."

We see that Einstein gives essentially three reasons for his original belief in assumption (ii), noting that, like two of the laws of thermodynamics, this assumption is a "principle of impotence," to use E. T. Whittaker's locution. Einstein's seemingly distinct three reasons are the following: (1) "On the basis of experience," there are no "stationary" light waves, (2) neither are there any such phenomena on the basis of Maxwell's equations, and (3) at the very beginning, there was intuitive clarity that preferred inertial systems do not exist, the laws of physics, including those of light propagation, being the same in all of them. These three reasons invite the following corresponding three comments.

1. The failure of our experience to have disclosed the existence

of stationary light waves is not, of course, presumptive of their nonexistence, unless that experience included the circumstances requisite to our observation of such waves if they do exist. What could such circumstances be? Suppose it *were* physically possible for a star to recede from the earth at the speed c of light. Assuming that there actually is such a star, postulate further that the speed of the light emitted by the star in the direction of the earth is c relatively to the *star*, the light's speed relatively to the earth being given by the *Galilean-Newtonian* velocity addition and hence being *zero*. Then the earth would maintain a *constant distance* from the light wave. And if there were a way for us to register the presence of that stationary light wave, then we could have evidence of its existence. *Mutatis mutandis*, a light source in the laboratory moving at the velocity c might have produced the same kind of phenomenon.

Perhaps Einstein envisaged these kinds of conditions as situations in which our experience ought to have disclosed the existence of stationary light waves.

If so, one wonders, however, how much weight he actually attached to this observational argument on behalf of assumption (ii). For he was undoubtedly cognizant of the contingency of the conditions governing the observable occurrence of the phenomenon in question. In particular, it should be noted that Einstein's mention of "the basis of experience" in this context *cannot* be assumed to be referring to the 1902–1906 experiments by Kaufmann and others on the deflection of electrons (β-rays) in electric and magnetic fields. For if we suppose him to have been familiar with these experiments, they must have left him in a quandary precisely in regard to the truth of assumption (ii): while yielding a mass-variation with velocity incompatible with *Newtonian* dynamics, the results of these experiments were unable to rule out the formulae of *Abraham's* dynamics, *which allowed* particle velocities *exceeding* the velocity of light *in vacuo*. . . .

2. Can stationary light waves be regarded as ruled out by Maxwell's equations, if one does not already accept the principle of relativity, which guarantees the validity of the usual form of Maxwell's equations in *all* inertial systems? In other words, can the *second* of Einstein's avowed reasons for his dismissal of the

possibility of stationary light waves be regarded as other than a logical consequence of the *third?* Clearly, *if* Maxwell's equations *are* coupled with the principle of relativity, then these equations indeed rule out stationary light waves in every inertial system. But since Maxwell's equations are not covariant under *Galilean* transformations, it is far from clear that stationary light waves are precluded by the form assumed by the equations in an inertial system S moving with the velocity c relatively to the primary (aether) frame K *and* having coordinates which are related by the *Galilean* transformations to those of the K system. Since Einstein does not mention the "Galilean transform" of Maxwell's equations, it would seem that the only reason why he felt justified in regarding Maxwell's equations as support for his repudiation of stationary light waves was that he had already assumed the principle of relativity on intuitive grounds.

3. In view of the presumably flimsy character of the appeal to experience and of the redundancy of (2) with (3) among the reasons given by Einstein, we are pretty much left with his intuitive confidence in the principle of relativity as the basis for his assumption of (ii).

We must emphatically *reject* the historical corollary of the not uncommon but altogether erroneous belief that the assertion of the limiting character of the velocity of light *in vacuo* depends on the relativistic velocity addition laws for its deduction. The following are compelling reasons for the *falsity* of this belief and hence of its historical corollary that Einstein arrived at assumption (ii) only after deducing the formulae for the composition of velocities.

He deduced the velocity addition laws in the fifth paragraph of his 1905 paper via the mediation of the Lorentz transformations from the two basic principles of his second paragraph—that is, from the principle of relativity and the principle of the constancy of the speed of light. Now, the latter principle presupposes, as he notes *pointedly* at the start of paragraph two, that the *one*-way transit-time ingredient in the *one*-way velocity of light is based on *the definition of simultaneity given in his first paragraph*. Since the deduction of the velocity addition laws thus presupposes the *denial* of absolute simultaneity, it clearly presupposes as *one* of its premisses the limiting character of the velocity of light *in vacuo*. Furthermore, the relativistic formulae for the composition of velocities show only that *if* each of the velocities to be added does *not* exceed *c*, then the addition of them will *not* result in a velocity greater than *c;* these formulae do *not* themselves show that velocities greater than *c* are physically impossible.

We are now in a position to make a conjecture as to Einstein's grounds for assumption (i). We recall that (i) asserts that within the class of events, material clocks do *not* define relations of absolute simultaneity under transport. As Einstein states explicitly both in our citation from his intellectual autobiography and in his formulation of the principle of the constancy of the speed of light in paragraph two of the 1905 paper, *the absolutistic conception of simultaneity* was the Gordian knot obstructing the resolution of his boyhood paradox—that is, the reconcilation of the two basic principles of his second paragraph. But, assumption (i) was *required* no less than (ii) for the denial of absolute simultaneity! Hence his confidence in (i) must be presumed to have derived from his belief in the correctness of *both* of the two principles in his second paragraph. . . .

Would it be safe to conclude from Einstein's autobiographical statement that actual experimental results in fact played no role at all when he groped his way to an espousal of the principle of relativity? If so, there would be the serious question whether the theoretical guesses of an Einstein can be regarded to have been genuinely more educated—as opposed to just more lucky—than the abortive phantasies of those quixotic scientific thinkers whose names have sunk into oblivion.

But is it true that actual experiments such as the Michelson-Morley experiment did *not* play any genetic role in the RT? E. T. Bell, writing *before* the publication of Einstein's "Autobiographical Notes," and referring to the influence of the Michelson-Morley experiment on Einstein, claims that "he has stated explicitly that he knew of neither the experiment nor its outcome when he had already convinced himself that the special theory was valid." And Mr. Polanyi reports that Einstein had authorized him in 1954 to publish the statement that "the Michelson-Morley experiment had a negligible effect on the discovery of relativity."

Yet if *both* of the reasons for RT adduced by Einstein in the "Introduction" to his 1905 paper were among the factors that had prompted his initial espousal of RT, the consonance of the foregoing claims by Bell and Polanyi with Einstein's own text is quite problematic. For in that "Introduction," Einstein cites the following two considerations as suggesting "that the phenomena of electrodynamics as well as of mechanics possess no properties corresponding to the idea of absolute rest": (1) The lack of *symmetry* in the classical electrodynamic treatment of a current-carrying wire moving relatively to a magnet at rest, on the one hand, and of a magnet moving relatively to such a wire at rest, on the other, *and* (2) "*the unsuccessful attempts to discover any motion of the earth relatively to the 'light medium'* [*aether*]." Unless they provide some other consistent explanation for the presence of the *latter* statement in Einstein's text of 1905, it is surely incumbent upon all those historians of RT who *deny* the inspirational role of the Michelson-Morley experiments to tell us *specifically* what *other* "unsuccessful attempts to discover any motion of the earth relatively to the 'light medium' " Einstein had in mind here. This obligation should also have been shouldered by the mature *reminiscing* Einstein himself when authorizing the statement given by Polanyi

PART IV

The Impact of Relativity Theory

THE SCIENTIFIC REACTION TO EINSTEIN

Relativity Theory had an unusual impact upon both the scientific and the lay public. For the scientist it came as a psychological shock of some magnitude to be told that many, if not most, of his cherished fundamental assumptions were wrong. Some made the transition to the new mechanics rather easily; for others, the change was a painful one, not always successfully managed.

For the layman the fuss, at first, seemed far removed from anything which could affect him. Yet as time went by it became clear that something of fundamental importance had taken place. Other areas of thought began to draw upon relativity, interpreting it in ways which must have made physicists squirm.

Both these reactions are illustrated in the selections which follow.

---

1 *William F. Magie* *"The Primary Concepts of Physics"*

On December 28, 1911, William F. Magie, Professor of Physics at Princeton University, delivered his presidential address before the American Physical Society in Washington, D. C. He there explained why he, for one, could not accept the basic postulates of Relativity Theory.

The subject of the present address is one that does not often appear on a scientific program. Physicists are so busy in enlarging

SOURCE. William F. Magie, "The Primary Concepts of Physics," *Science*, New Series, Vol. XXXV, January-June 1912, pp. 281, 287, and 290–293. Reprinted by permission of the publisher.

the structure of knowledge that few of them concern themselves with the consideration of the fundamental concepts of the science. Yet it is plainly true that if those fundamental or primary concepts are not clearly apprehended, or if there is doubt as to what they are, the whole structure of the science rests on an insecure basis. I propose to examine certain questions concerning these primary concepts, about which there has been and is much unsettled opinion. . . .

I now come to a much more difficult part of my subject, the consideration of the other primary concepts of space and time. Not many years ago we should have been willing to pass them over with a mere mention, admitting the impossibility of giving a definition or even an intelligible description of either of them, admitting the impossibility of determining an absolute or fixed point in space, or an absolute instant of time, but still asserting that we knew something about them both of which we were sure. At present we are driven by the development of the principle of relativity to examine anew the foundations of our thought in respect to these two primary concepts. . . .

The principle of relativity in this metaphysical form professes to be able to abandon the hypothesis of an ether. All the necessary descriptions of the crucial experiments in optics and electricity by which the theories of the universe are now being tested can be given without the use of that hypothesis. Indeed the principle asserts our inability even to determine any one frame of reference that can be distinguished from another, or, what means the same thing, to detect any relative motion of the earth and the ether, and so to ascribe to the ether any sort of motion; from which it is concluded that the philosophical course is to abandon the concept of the ether altogether. This question will be amply and ably discussed this morning, but I may venture to say that in my opinion the abandonment of the hypothesis of an ether at the present time is a great and serious retrograde step in the development of speculative physics. The principle of relativity accounts for the negative result of the experiment of Michelson and Morley, but without an ether how do we account for the interference phenomena which made that experiment possible? There are only two ways yet thought of to account for the passage of light through space. Are the supporters of the theory of relativity going to return to the corpuscles of Newton? Are they willing to explain the colors of thin plates by invoking "the fits of easy reflection and of easy transmission?" Are they satisfied to say about diffraction that the corpuscles near an obstacle "move backwards and forwards with a motion like that of an eel"? How are they going to explain the plain facts of optics? Presumably they are postponing this necessary business until the consequences of the principle of relativity have been worked out. Perhaps there is some other conceivable mode of connection between bodies, by means of which periodic disturbances can be transmitted. We may imagine a sort of tentacular ether stretching like strings from electron to electron, serving as physical lines of force, and transmitting waves as a vibrating string does. Such a luminiferous medium would not meet the postulate of simplicity, but it conceivably might work. But whatever the properties of the medium may be, there is choice only between corpuscles and a medium, and I submit that it is incumbent upon the advocates of the new views to propose and develop an explanation of the transmission of light and of the phenomena which have been interpreted for so long as demonstrating its periodicity. Otherwise they are asking us to abandon what has furnished a sound basis for the interpretation of phenomena and for constructive work in order to preserve the universality of a metaphysical postulate. . . .

But, after all, these questions raised by the development of the principle of relativity are of secondary importance. The central question is whether or not this principle can ever furnish a satisfactory explanation of natural phenomena. The formulas derived from it are evidently merely descriptive. . . .

We can understand from what we see and feel what is meant by the motions of elastic spheres, and the model which uses them to represent the behavior of a gas is not only competent to reproduce the behavior of a gas, but is intelligible in the elements of which it is composed. The model of the elastic solid ether, incomplete and objectionable as it became when the subject of optics was enlarged and developed, was intelligible in its elements. The model of electromagnetic operations embodied in Maxwell's formulas is also one which is thus intelligible in its elements. When I say this I do not mean that we know all about electric and magnetic forces, but I mean that we do know enough about such forces to have a clear notion of their variation in space and their variation in time.

This feature of the ideal model or description seems to me to be necessary in order to make the model acceptable as the ultimate or last attainable explanation of phenomena. The elements of which the model is constructed must be of types which are immediately perceived by the senses and which are accepted by everybody as the ultimate data of consciousness. It is only out of such elements that an explanation, in distinction from a mere barren set of formulas, can be constructed. A description of phenomena in terms of four dimensions in space would be unsatisfactory to me as an explanation, because by no stretch of my imagination can I make myself believe in the reality of a fourth dimension. The description of phenomena in terms of a time which is a function of the velocity of the body on which I reside will be, I fear, equally unsatisfactory to me, because, try I ever so hard, I can not make myself realize that such a time is conceivable.

Tried by this test, I feel that the principle of relativity does not speak the final word in the discussion about the structure of the universe. The formulas which flow from it may be in complete accord with all discovered truth, but they are expressed in terms which themselves are not in harmony with my ultimate notions about space and time. That this is true is so evident that it is generally admitted. . . .

Therefore, from my point of view, I can not see in the principle of relativity the ultimate solution of the problem of the universe. A solution to be really serviceable must be intelligible to everybody, to the common man as well as to the trained scholar. All previous physical theories have been thus intelligible. Can we venture to believe that the new space and time introduced by the principle of relativity are either thus intelligible now or will become so hereafter? A theory becomes intelligible when it is expressed in terms of the primary concepts of force, space and time, as they are understood by the whole race of man. When a physical law is expressed in terms of those concepts we feel that we have a reason for it, we rest intellectually satisfied on the ultimate basis of immediate knowledge. Have we not a right to ask of those leaders of thought to whom we owe the development of the theory of relativity, that they recognize the limited and partial applicability of that theory and its inability to describe the universe in intelligible terms, and to exhort them to pursue their brilliant course until they succeed in explaining the principle of relativity by reducing it to a mode of action expressed in terms of the primary concepts of physics?

2 *Louis Trenchard More* *"The Theory of Relativity"*

Louis Trenchard More, himself a physicist, reported the speech of Professor Magie to the readers of The Nation. *His objections to the new Relativity Theory emerge quite clearly from what he considered science itself to be.*

The recent joint meeting of the American Physical Society and Section B of the American Association for the Advancement of Science, held at Washington in the Christmas holidays, was distinguished chiefly by the attention given to certain novel theories of physics. The presidential address of Prof. W. F. Magie, "On the Primary Concepts of Physics," is significant of a reaction against the domination of these new ideas; and the symposium which followed showed very clearly a sharp line of cleavage between the classical and modern schools of mechanics. . . .

The science of physics seems to be suffering, these latter days, from an attack of intellectual indigestion. While physicists feel that their subject has always shown a healthy growth, yet, as a rule, new discoveries have been made slowly enough to be fitted into theory without causing serious trouble. . . . Since the begin-

SOURCE. Louis Trenchard More, "The Theory of Relativity," *The Nation*, Vol. XCIV, January-June 1912, pp. 370–371. Reprinted by permission of the publisher.

ning of modern physics, from the days, that is, of Galileo and Newton, physicists have been building their laws and their theories on the same primary mechanical concepts of space, time, and mass. Through all this time, the first two have evoked little discussion, and differences of opinion about the concept of matter have been, for the most part, merely a question of precedence regarding mass, force, and energy. . . .

Now, in the past, as new phenomena were discovered, theories were advanced to explain them in terms of these primary mechanical concepts, and if discrepancies remained between the theory and the phenomena, the theory was abandoned or allowed to lie dormant, but the concepts were not questioned. This may be called the classic attitude; but a new scientific method, which may be called the school of transcendental symbolism, has been lately evolved by German physicists. . . .

Professor Einstein takes, as his starting point, the fact that certain experiments (especially one by Professors Michelson and Morley), to determine the mutual action of matter and the æther on the velocity of light, fail to give any positive results. He therefore accepts this nugatory result; assumes, as a postulate, that the velocity of light in space is an absolute constant unaffected by the motion of matter, in conformity with the experiment, and denies the existence of the æther. From thence, by steps we need not here follow, he also draws the conclusion that we must radically alter our concepts of space and time, and abandon our concept of mass. In this new Theory of Relativity, as it is called, the dimensions and the inertia of a body, and the measurement of time, are not stationary quantities, but vary in accordance with the velocity of the body as it moves. Furthermore this relativity of mass and time to motion depends on a mathematical formula purely abstract in source and character. This really amounts to saying that experience is not a criterion of truth and that we must rely on an inward sentiment of knowledge as revealed in subjective formulæ.

Both Professor Einstein's theory of Relativity and Professor Planck's theory of Quanta are proclaimed somewhat noisily to be the greatest revolutions in scientific method since the time of Newton. That they are revolutionary there can be no doubt, in so far as they substitute mathematical symbols as the basis of science and deny that any concrete experience underlies these symbols, thus replacing an objective by a subjective universe. The question remains whether this change is a step forward or backward, into light or into obscurity. It is held, and apparently rightly, that the revolution effected by Galileo and Newton was to replace the metaphysical methods of the schoolman by the experimental methods of the scientist. Now the new methods might seem to be just the reversal of that step, so that, if there is here any revolution in thought, it is in reality a return to the scholastic methods of the Middle Ages.

Undoubtedly the German mind is prone to carry a theory to its logical conclusion, even if it leads into unfathomable depths. On the other hand, Anglo-Saxons are apt to demand a practical result, even at the expense of logic. . . .

When Professor Magie undertakes the consideration of the other two primary concepts, space and time, one notices a symptom of uncertainty and restlessness. Past writers discussed these concepts very briefly, whereas those who are now advocating the Principle of Relativity do so with such an air of assurance and finality that a modest man may hesitate to express his doubts. They may stagger us when they require us to believe that the length of a body becomes less if it is put in motion, and that clocks run slower when they move than when they are at rest; but, on the other hand, they offer the most alluring seduction to the mind, when, by the simplest kind of mathematics, they appear to subdue the whole universe to their ideas. Professor Magie points out that the chief incentive to the development of the theory of relativity is the desire to express all natural phenomena by a set of simple equations; and he is right when he objects to making the demand for simplicity the chief purpose of a scientific theory. It is better to keep science in homely contact with our sensations at the expense of unity than to build a universe on a simplified scheme of abstract equations. The main question, however, is whether or not the principle will explain natural phenomena in a satisfactory manner as they appear to us. Professor Magie evidently thinks it will not, and that we had better keep to the concrete models of atoms and the æther, which are imaginable even if they are quite artificial. And in the last analysis, a solution of our problems must be intelligible to the man of general intelligence as well as to the trained specialist. From the contradictory statements of the specialists themselves he might also include them in the class which finds the Principle of Relativity of dubious clarity.

3 FROM *Arthur Stanley Eddington* *The Theory of Relativity and its Influence on Scientific Thought*

To those scientists who could free themselves from their own preconceptions, the Einsteinian view of the universe, through Relativity Theory, was nothing less than exhilarating. Arthur Stanley Eddington was particularly well placed to appreciate the new perspective. It was he who had led the 1919 eclipse expedition from Great Britain, which confirmed Einstein's prediction that the light from a star would be bent as it passed through the sun's gravitational field. The very subjectivity of which More complained was part of the appeal of Relativity Theory to Eddington. Furthermore, the fact that the destruction of absolute space and time might lead to a resurgence of "mysticism" was not altogether distasteful to Eddington. A "mysticism" that had the backing of mathematical physics might well be worth exploring! Was not the ancient dream of philosophy the union of man and the universe?

In the days before Copernicus the earth was, so it seemed, an immovable foundation on which the whole structure of the heavens was reared. Man, favourably situated at the hub of the universe, might well expect that to him the scheme of nature would unfold itself in its simplest aspect. But the behaviour of

SOURCE. Arthur Stanley Eddington, *The Theory of Relativity and Its Influence on Scientific Thought* (The Romanes Lecture 1922), Oxford: The Clarendon Press, 1922, pp. 3–6, 11–12, and 31–32. Reprinted by permission of the publisher.

the heavenly bodies was not at all simple; and the planets literally looped the loop in fantastic curves called epicycles. The cosmogonist had to fill the skies with spheres revolving upon spheres to bear the planets in their appointed orbits; and wheels were added to wheels until the music of the spheres seemed wellnigh drowned in a discord of whirling machinery. Then came one of the great revolutions of scientific thought, which swept aside the Ptolemaic system of spheres and epicycles, and revealed the simple plan of the solar system which has endured to this day.

The revolution consisted in changing the view-point from which the phenomena were regarded. As presented to the earth the track of a planet is an elaborate epicycle; but Copernicus bade us transfer ourselves to the sun and look again. Instead of a path with loops and nodes, the orbit is now seen to be one of the most elementary curves—an ellipse. We have to realize that the little planet on which we stand is of no great account in the general scheme of nature; to unravel that scheme we must first disembarrass nature of the distortions arising from the local point of view from which we observe it. The sun, not the earth, is the real centre of the scheme of things—at least of those things in which astronomers at that time had interested themselves—and by transferring our view-point to the sun the simplicity of the planetary system becomes apparent. The need for a cumbrous machinery of spheres and wheels has disappeared.

Every one now admits that the Ptolemaic system, which regarded the earth as the centre of all things belongs to the dark ages. But to our dismay we have discovered that the same *geocentric* outlook still permeates modern physics through and through, unsuspected until recently. It has been left to Einstein to carry forward the revolution begun by Copernicus—to free our conception of nature from the terrestrial bias imported into it by the limitations of our earthbound experience. To achieve a more neutral point of view we have to imagine a visit to some other heavenly body. That is a theme which has attracted the popular novelist, and we often smile at his mistakes when sooner or later he forgets where he is supposed to be and endows his voyagers with some purely terrestrial appanage impossible on the star they are visiting. But scientific men, who have not the novelist's licence, have made the same blunder. When, following Copernicus, they station themselves on the sun, they do not realize that they must leave behind a certain purely terrestrial appanage, namely, *the frame of space and time* in which men on this earth are accustomed to locate the events that happen. It is true that the observer on the sun will still locate his experiences in a frame of space and time, if he uses the same faculties of perception and the same methods of scientific measurement as on the earth; but the solar frame of space and time is not precisely the same as the terrestrial frame, as we shall presently see. . . .

The more closely we examine the processes by which events are assigned to their positions in space and time, the more clearly do we see that our local circumstances play a considerable part in it. We have no more right to expect that the space-time frame on the sun will be identical with our frame on the earth than to expect that the force of gravity will be the same there as here. If there were no experimental evidence in support of Einstein's theory, it would nevertheless have made a notable advance by exposing a fallacy underlying the older mode of thought—the fallacy of attributing unquestioningly a more than local significance to our terrestrial reckoning of space and time. But there is abundant experimental evidence for detecting and determining the difference between the frames of differently circumstanced observers. Much of the evidence is too technical to be discussed here, and I can only refer to the Michelson-Morley experiment. I fear that some of you must be getting rather tired of the Michelson-Morley experiment; but those who go to a performance of Hamlet have to put up with the Prince of Denmark. . . .

It is sometimes complained that Einstein's conclusion that the frame of space and time is different for observers with different motions tends to make a mystery of a phenomenon which is not after all intrinsically strange. We have seen that it depends on a contraction of moving objects which turns out to be quite in accordance with Maxwell's classical theory. But even if we have succeeded in explaining it to ourselves intelligibly, that does not make the statement any the less true! A new result may often be expressed in various ways; one mode of statement may sound less mysterious; but another mode may show more clearly what will be the consequences in amending and extending our knowledge. It is for the latter reason that we emphasize the relativity of space—that lengths and distances differ according to the observer implied. Distance and duration are the most fundamental terms in physics; velocity, acceleration, force, energy, and so on, all depend on them; and we can scarcely make any statement in physics without direct or indirect reference to them. Surely then we can best indicate the revolutionary consequences of what we have learnt by the statement that distance and duration, and all the physical quantities derived from them, do not as hitherto supposed refer to anything absolute in the external world, but are relative quantities which alter when we pass from one observer to another with different motion. The consequence in physics of the discovery that a yard is not an absolute chunk of space, and that what is a yard for one observer may be eighteen inches for another observer, may be compared with the consequences in economics of the discovery that a pound sterling is not an absolute quantity of wealth, and in certain circumstances may "really" be seven and sixpence. The theorist may complain that this last statement tends to make a mystery of phenomena of currency which have really an intelligible explanation; but it is a statement which commends itself to the man who has an eye to the practical applications of currency.

Ptolemy on the earth and Copernicus on the sun are both contemplating the same external universe. But their experiences are different, and it is in the process of experiencing events that they become fitted into the frame of space and time—the frame being different according to the local circumstances of the observer who is experiencing them. That, I take it, is Kant's doctrine, "Space and time are forms of experience." The frame then is not in the world; it is supplied by the observer and depends on him. And those relations of simplicity, which we seek when we try to obtain a comprehension of how the universe functions, must lie in the events themselves before they have been arbitrarily fitted into the frame. The most we can hope for from any frame is that it will not have distorted the simplicity which was originally present; whilst an ill-chosen frame may play havoc with the natural simplicity of things. . . .

If I have succeeded in my object, you will have realized that the present revolution of scientific thought follows in natural sequence on the great revolutions at earlier epochs in the history of science. Einstein's special theory of relativity, which explains the indeterminateness of the frame of space and time, crowns the work of Copernicus who first led us to give up our insistence on a geocentric outlook on nature; Einstein's general theory of relativity, which reveals the curvature of non-Euclidean geometry of space and time, carries forward the rudimentary thought of those earlier astronomers who first contemplated the possibility that their existence lay on something which was not flat. These earlier revolutions are still a source of perplexity in childhood, which we soon outgrow; and a time will come when Einstein's amazing revelations have likewise sunk into the commonplaces of educated thought.

To free our thought from the fetters of space and time is an aspiration of the poet and the mystic, viewed somewhat coldly by the scientist who has too good reason to fear the confusion of loose ideas likely to ensue. If others have had a suspicion of the end to be desired, it has been left to Einstein to show the way to rid ourselves of these "terrestrial adhesions to thought." And in removing our fetters he leaves us, not (as might have been feared) vague generalities for the ecstatic contemplation of the mystic, but a precise scheme of world-structure to engage the mathematical physicist.

THE LAY REACTION

4 EDITORIAL *The New York Times* *"A Mystic Universe" January 28, 1928*

The combination of Relativity Theory and the new Wave Mechanics was too much for the New York Times. *In an editorial of January 28, 1928, the* Times *expressed what must have been the general reaction of many lay people who tried to keep up with the new physics but found its content totally bewildering.*

Tennyson claimed for faith the function of believing what we cannot prove. The new physics comes perilously close to proving what most of us cannot believe; at least until we have rid ourselves completely of established notions and forms of thought. Relativity translates time into terms of space and space into terms of time. The Quantum invites us to think of something which can be in two places at the same time, or which can move from one spot to another without passing through intervening space. Matter, which is hard enough to grasp as electronic activity, becomes still harder to visualize as mere pulsations. It can scarcely be called "matter," this physical substratum of being which has been rarefied to a degree where matter threatens to blend into what used to be called spirit.

Not even the old and much simpler Newtonian physics was comprehensible to the man in the street. To understand the new

SOURCE. *The New York Times,* January 28, 1928, p. 14, col. 3.

physics is apparently given only to the highest flight of mathematicians. Countless textbooks on Relativity have made a brave try at explaining and have succeeded at most in conveying a vague sense of analogy or metaphor, dimly perceptible while one follows the argument painfully word by word and lost when one lifts his mind from the text. It is a rare exposition of Relativity that does not find it necessary to warn the reader that here and here and here he had better not try to understand. Understanding the new physics is like the new physical universe itself. We cannot grasp it by sequential thinking. We can only hope for dim enlightenment.

The situation is all the harder on the public because physics has become unintelligible precisely in an age when the citizen is supposed to be under the moral obligation to try to understand everything. Nor are things made easier for the common man—meaning the man untrained in the highest mathematics—when theory changes from year to year. The validity of the Niels Bohr atom of positive and negative particles of electricity has just been questioned by a speaker before the New York Electrical Society, who substitutes for electric particles groups of whirling waves consisting of "nobody knows what." You have on the one hand X-rays identified as showers of particles and atomic particles conjectured as rays or waves. The ether, which has gone out of fashion in recent years, is now apparently by way of being rehabilitated.

In this turmoil there is at least one possible source of comfort. Earnest people who have considered it their duty to keep abreast of science by readapting their lives to the new physics may now safely wait until the results of the new discoveries have been fully tested out by time, harmonized and sifted down to a formula that will hold for a fair term of years. It would be a pity to develop an electronic marriage morality and find that the universe is after all ether, or to develop a wave code for fathers and children only to have it turn out that the family is determined not by waves but by particles. Arduous enough is the task of trying to understand the new physics, but there is no harm in trying. Reshaping life in accordance with the new physics is no use at all. Much better to wait for the new physics to reshape our lives for us as the Newtonian science did.

5 *H. Wildon Carr*
"Metaphysics and Materialism"

For the average man, Relativity Theory, as such, was as remote as the peculiar action of light passing through the sun's gravitational field. But as the history of science had amply shown in the past, today's physical theory can often affect tomorrow's daily life. In the 1920's, God was not yet declared dead and there were those, like Professor Carr in the following selection, who felt that the theological aspects of Relativity Theory should not be neglected. The riposte by Mr. Hugh Elliot illustrates the power of Relativity Theory to fit different intellectual frames of reference. Both the article by Carr and the answer by Eliot appeared in the British journal Nature, *one of the most prestigious scientific journals.*

If the illusion of the scholastic method is that from mere forms we can deduce essences, then the world-view which we call materialism is only a scholastic pastime." This is the concluding sentence of Hermann Weyl's "Raum, Zeit, Materie." Whatever may be the case with the physicists, the mathematicians are under no illusion with regard to the completeness of the scientific revolution. The principle of relativity has not merely complicated the concept of physical reality; it has re-formed it. Mathematics is, and has always been recognized as being, a constructive process of the human mind exercised on physical existence. The old mathematics took its matter from physics; the new mathematics gives matter to physics. The effect is that the world-view which had become for physical science in the nineteenth century practically unchallengeable, and the acceptance of which had come

SOURCE. H. Wildon Carr, "Metaphysics and Materialism," *Nature,* Vol. CVIII, October 20, 1921, pp. 247–248. Reprinted by permission of The Macmillan Company.

to be regarded as the indispensable condition and only passport for those who would enter the ranks of scientific investigators, has become suddenly incredible. It is true, indeed, that it still has its defenders, and that it is held as firmly as ever by many who continue to be in their special departments authoritative teachers; but this does not alter the fact that for us to-day the world-view is changed, and it is not even strange that many leaders in scientific research still cling fast to the old view when we remember that the great originator of the modern inductive method in the seventeenth century, Francis Bacon, to the end rejected the Copernican theory.

Materialism does not stand for any particular theory of the nature of matter, but for the general world-view that matter, something *de facto* objective the ultimate constitution of which we may not know, and even may not be able to know, but which is entirely independent of our reason and of any thoughts we may have about it, exists and constitutes the reality of the universe, including reason and will, which as qualities or properties of some of its forms give rise to knowledge of it. . . .

It may not be obvious at once that the mere rejection of the Newtonian concept of absolute space and time and the substitution of Einstein's space-time is the death-knell of materialism, but reflection will show that it must be so. If space is not endless, but finite (and this is the essential principle of the Riemannian geometry), and if time is not in its existence independent of space, but co-ordinate with the spatial dimensions in the space-time system (and this is the essential principle of the concept of the four-dimensional continuum), then the very foundation of the materialistic concept is undermined. For the concept of relative space-time systems the existence of mind is essential. To use the language of philosophy, mind is an *a priori* condition of the possibility of space-time systems; without it they not only lose meaning, but also lack any basis of existence. The co-ordinations presuppose the activity of an observer and enter into the constitution of his mind. If you distinguish, as, of course, you must and do, the observer from his space-time system, it is not a distinction of two separate existences externally related; they exist only in their relation, as when, for example, we distinguish an activity from its expression. . . .

Materialism is essentially a monistic and atomistic conception of reality. For it matter is primordial, and mind is derived. Philosophers from the beginning of philosophy have been conscious of the intellectual difficulty of such a concept, but it has always seemed, even to philosophers, a necessary presupposition of physical science. Science, it was conceded, must at least proceed *as if* it were so. The principle of relativity is the rejection of it, a rejection based on the discovery, not of theoretical difficulties, but of practical matters of fact. The supposed fundamental reality on which materialism as a world-view was supported has proved a vain illusion, and materialism is left in the air. The new scientific conception of the universe is monadic. The concrete unit of scientific reality is not an indivisible particle adversely occupying space and unchanging throughout time, but a system of reference the active centre of which is an observer co-ordinating his universe. The methodological difference between the old and the new is that mathematics is a material, and no longer a purely formal, science.

6 *Hugh Elliot*
"Relativity and Materialism"

Prof. Wildon Carr has for a number of years been busily engaged in ringing the death-knell of materialism. I was therefore not a little surprised to read in *Nature* (October 20) his statement that Einstein's theory was the "death-knell of materialism." I thought, from my previous acquaintance with Prof. Carr's writings, that Bergson, Croce, and others had already done all that was necessary in that direction. But no! Prof. Carr has resuscitated it for the express purpose of killing it once more. That unfortunate doctrine seems to exist mainly for the purpose of being periodically slaughtered by professors of metaphysics; and we are led to the conviction that materialism must have very

SOURCE. Hugh Elliott, "Relativity and Materialism," *Nature*, Vol. CVIII, December 1, 1921, p. 432. Reprinted by permission of The Macmillan Company.

singular properties to survive so many tragic executions.

Well, it does possess a property which must naturally appear singular to those steeped in metaphysics–it happens to be true. Scientific materialism, as now understood, does not profess to be a rounded or final system of philosophy: it is merely a name for a few general principles, laid down by science, and selected for emphasis on account of their high human significance. Science makes new knowledge; philosophy (rightly understood) does not; it simply collects together certain principles yielded by science, those principles being selected as having some bearing on the deep undying problems of most profound human interest.

Among the scientific principles thus selected and emphasised by materialism–and the only one among them still seriously controverted–is that which states that mind cannot exist apart from matter, or as I prefer to put it, that mind is a function of material organisms. Prof. Wildon Carr is of opinion that mind *can* and *does* exist apart from matter; and he is under the impression that this opinion is justified by the principle of relativity. So far as I can follow his argument, it amounts to this. Space and time are relative to the observer; therefore the existence of an observing mind must be antecedent to the existence of space and time. True; but space and time are not matter: they are not objective things; you cannot weigh them or touch them; they are part of the mental framework which we erect for our convenience in dealing with external nature. They are concepts; just as the number 10 is a concept; not a thing, but a framework into which things can be fitted. "For the concept of relative space-time systems the existence of mind is essential." Prof. Carr might with equal profundity have said that for the presence of dew the existence of water is essential. Dew is aqueous; a concept is mental; but let me inform Prof. Carr that neither one nor the other of these propositions gives the slightest qualm to any scientific materialist, nor have they the least relevance to the question whether or not mind depends upon matter. We are not concerned with "concepts," which, of course, imply the previous existence of mind, but with objective *things*.

Now Prof. Carr argues that the "space-time system," involved by relativity, is conditional on the existence of mind. Very well then. It follows that if mind were to be extinguished throughout the universe, the laws at present ascribed to the universe would cease to operate, or perhaps the universe itself would cease to exist. Now that is an altogether incredible proposition. If Prof. Carr's mind were to be extinguished, the laws of nature would still remain as they are. If everybody else's mind were also to be extingiushed the laws of nature would be unaltered. "Concepts" would vanish no doubt; but the validity of the principle of relativity itself does not depend on the existence of a mind which can testify to it. Prof. Carr exhibits that incurable confusion between concepts and objects which is common to all those who think that metaphysics is a rival method of science in the making of new knowledge.

Relativity of space and time no more conflicts with scientific materialism than does relativity of motion. But it is idle to argue with sentiment, and it is with sentiment alone that we have to do—sentiment unsupported by a fragment of evidence, and asserting itself in flat contradiction to every principle of logic. As a mere statement of truth, materialism will always reign, as it has reigned now for centuries as the basis of scientific experiment. But on a show of hands it will always be in a minority; its reign is that of an uncrowned king. There exists a wide and universal human sentiment which loathes materialism. That sentiment comes out in many different forms: in the vulgar superstitions of the uneducated, in spiritualism, in metaphysical dissertation. They are but the same deep sentiment on different intellectual grades, but as false and rotten in the higher grades as they are in the lower. Everywhere it comes out: in physiology we find it as vitalism; among the public at large it supports religion, the most powerful single factor that has moulded the destinies of civilised humanity. Materialism must always be unpopular; that is why it is so often being killed. But it is true; that is why it never dies; that is why it never will die; unless, indeed, it is one day drowned in the floods of oily sentimentalism.

The advent of Relativity Theory was often hailed by the non-scientist involved in other areas of creative activity as a liberating influence. The number of articles or books that appeared in the 1920's and 1930's using Einstein's views to support heterodox opinions or movements is literally legion. Two examples must suffice to illustrate this aspect of the impact of Relativity Theory.

In the first Relativity Theory is related to art by Thomas Jewell Craven, historian and critic of painting. In the second psychology is examined in the light of Relativity.

---

7 *Thomas Jewell Craven* *"Art and Relativity"*

Professor Einstein's revolutionary theory is the latest example of the eternal kinship between art and science. His principle of relativity, essentially valid in the unbounded realm of mechanics, leads portentously to an aesthetic analogue which has hitherto received no critical attention. It has long been recognized in the plastic arts that the potentialities of linear alteration are governed by the design, a fact as familiar to the psychologist as to the painter; the relativity of colour values is equally well known, but this interdependence, because of its endless range, has never been fully catalogued. While the celebrated physicist has been evolving his shocking theories of the courses of natural phenomena, the world of art has suffered an equivalent heterodoxy with respect to its expressive media. This revolt has sprung from the conviction that the old art is not necessarily infallible, and that equally significant achievements may be reached by new processes and by fresh sources of inspiration. . . .

In his special theory of relativity Professor Einstein has demonstrated with brilliant finality that Newton's laws of inertia are true only for a Newtonian system of co-ordinates; that is, when the gravitational field is disregarded, and when the description of motions is definitely referable to a point on a rigid body of

SOURCE. Thomas Jewell Craven, "Art and Relativity," *The Dial,* Vol. LXX, January-June 1921, pp. 535–539.

specification; he has shown that these laws are adequate for practical measurements but incompatible with the law of the propagation of light unless the Lorentz transformation be substituted. In his general theory he has defined the limited validity of the special principle, and has made clear that the laws of natural phenomena cannot be formulated with absolute accuracy unless the old co-ordinates are abolished and a new system devised wherein the reference-bodies are no longer fixed but in relative motion. In connecting the equations of an abstract science like mathematics with philosophy the symbolical method must be followed; in the case of art the same plan is retained, and with even more striking results. When one considers the reflective aspects of art and its close affinity with the general thought of its time, this connection will not seem strange. The plastic world is, of course, compounded of manifold details gathered from the forms of perceptional experience, but the processes involved in harmonizing these details are purely psychic and inseparably bound to all other psychic factors of the age. It is hardly necessary to add that neither scientific nor mathematical formulae are directly concerned with this reaction to life, and that the quest for new relations in art-forms is guided almost entirely by feeling after the first intellectual step has been taken.

The fixed co-ordinates upon which the Newtonian measurements were erected have their parallel in more than one aesthetic manifestation. It is of no consequence that these manifestations have differed in tendency—there has always existed a common bond of interest, a rigid system of judgements corresponding to an immovable reference-body, and it is this abstract quality which establishes the analogy between the old art and classical mechanics. Professor Einstein's general theory of relativity has taken the whole physical structure; similarly has the modern painter broken the classical traditions.

Although the artists of the past, in striving for enduring beauty, never regarded organization as an end, nevertheless they were conscious of its importance; and in every period the creative will has received its impetus from specific and rigid tenets. Most of these principles since the days of Giotto have been founded upon verisimilitude, architectural proportion, and the like; they have been born of the belief that truth could not be attained except by strict adherence to the dictates of experience. Co-ordinates from which further relations were constructed have varied from time to time, but in every movement to the present they have had inception in inflexible ideas, such as the logic of light and shade, correct anatomical structure, and perspective. Even in rhythm the balancing actions and counter-actions have become standardized, and composition has deteriorated into mechanical pattern-making. The artist of to-day is not seeking the impossible, the overthrow of the past; he asks that the relativity of individual truths be acknowledged; he is convinced that the real meaning of art lies beyond precise lines of definition, and is searching for a new point of departure, a system of co-ordinates which allows him to achieve coherence without falling back on the laws of visual experience, knowing that these laws invariably become static and conventionalized when severed from the field of personal action where they originate. It is undeniable that the great man of former periods has broken the laws of his age, has revolted against the aesthetic dogma handed down to him; but what has signalized his genius has not been the construction of a new and moving reference-body, but a change in direction from a fixed basis.

It is at last recognized that the truth of art from a constructive point of view is a matter of coherence, of inevitable relationships, and that to intensify its value as a reflection of life, art can no longer proceed from the traditional loci. Instead of clinging to the rigid laws of photographic vision for a logic of creative activity, the modernist is ever mindful of his psychic responses to experience. For example: a painter has chosen for a theme a specific landscape consisting, say, of two houses, a prominent tree, a brook, and a bridge, items which may be delineated in several ways, and which may be held together pictorially by following a precise scheme of light and shade, by obedience to correct perspective, or by certain recurrent accents of lines. Each of these methods is compatible with the old doctrine of art, and each is adequate for graphic rendition; but it is not to be inferred that the primary inspiration of the painter was the simple idea of representation. What made it his own theme was the fact that the landscape aroused his perceptive powers and stirred his emotions—it had characteristics peculiar to him alone. It is here that a factor enters the old system of co-ordinates which is quite as disturbing as Professor Einstein's introduction of the time element into the Euclidean laws of spatial calculation. The personal feeling of the artist must be injected to arrive at greater truth, a truth far beyond that of mere vision, for the latter quality, while it serves all purposes of illustration, reveals nothing psychologically.

In the landscape mentioned above, the painter feels the predominance of certain forms; some objects attract him and stimulate his imagination profoundly—others are instinctively allotted a secondary position; the forms which excite him are contemplated, one might say, out of perspective, out of the pure logic of vision—they assume a magnitude that transcends all reality. Obviously the artist's conception of the real and living truth cannot be portrayed by conformity to any laws of actual appearance—it is compassed in a different fashion. Nor can the goal be reached by the simple device of accentuations, for here he is confronted with the fundamental requisite of coherence which insists on the relativity of the constituent parts in spite of all emphasis. He must, therefore, discover some point of reference that will provide for the desired accentuations and at the same time preserve unity and sequence without which art is inconceivable.

It is here that organization becomes a decidedly conscious process, and proclaims the necessity for a new and mobile basis identified with the personal element. We must not conclude that such an element has been absent in the old art; but not until modern times has painting been regarded as a vehicle for psychological truth, has it been made the reflection of the artist's mental states in the presence of simple objects of experience. The message of the former periods, notably in the great ages of productivity, has been spiritual in the collective sense—pervaded with religious thought; to-day it testifies to individual psychology and mirrors scientific experiment. Seizing the old system of visual coordinates, the modern painter has infused into it the personal element with a high degree of premeditation, and in place of the static pivot, correct in architectural symmetry, sound in aerial perspective, and logical in light and shade, he has given us a moving body of specification, independent of naturalism of any sort, and by which the integral forms are bound together by flowing sequences of line and colour.

Recognizable objects find their way as often as not into the new works of art, but they are never servile to realistic appearance, and it is unlikely that the painter will ever again attempt the ancient efforts to reproduce nature literally. Endowed with the system of co-ordinates gradually evolved since the death of Cézanne he has at his command the most plastic medium of expression that the world has ever known.

8 *Paul Chatham Squires*
"A New Psychology after the Manner of Einstein"

I

In an era marked by such intense interest concerning things psychological as is the present one, many aspects of psychology have found expression in the popular literature. Especially have the problems of sex and the occult found a dominant place in the public mind. But there is a most fascinating chapter in psychology, hitherto unrevealed to the layman, that affords a striking comparison of a new movement in psychology with the principle of relativity made current and popular by the great physicist Einstein. . . .

The Gestalt psychology, arising in Germany within the course of the past two decades, has been enthusiastically heralded by some as furnishing the one and only adequate key to the problems of mind, while by others it has been made the target of violent criticism. Whatever the final verdict of science in respect to the claims of this new school of psychology may be, configurationism has at any rate succeeded brilliantly in agitating and

SOURCE. Paul Chatham Squires, "A New Psychology after the Manner of Einstein," *The Scientific Monthly*, Vol. XXX, January-June 1930, pp. 156–157 and 159–163. Reprinted by permission of the publisher.

redirecting the currents of traditional thought. For just as Einstein gave an impetus to physics by expounding the relative nature of space and time, so the champions of the configuration psychology have been assiduously engaged in the attempt to demonstrate the relative character of our mental life, and have thereby imbued present-day psychology at large with renewed vigor.

Every one has a practical understanding of the word configuration. It means the outline, the shape, the contour of an object; thus, we may speak of the configuration of a human face or the configuration of a mountain. And when we are studying the configuration of something we are either ignoring or treating as unimportant for our present purposes the elements, the bricks as it were, which enter into the composition of the given structure.

The use of the term Gestalt or configuration, then, implies that our attention is focused upon the object-considered-as-a-whole, and not upon the object-considered-as-consisting-of-parts. Here you have unearthed the fundamental clue to the secret of the configuration psychology. For this new psychology interests itself primarily, if not exclusively, with the form, rather than with the so-called elements, of mental activity. The older psychology, however, was preoccupied with a highly abstract dissection of mind into its ultimate parts, and hence strongly resembled an atlas of anatomy. Toward this anatomical attitude the configuration psychology is utterly hostile. . . .

II

Let us make the acquaintance of the new relativity psychology by means of a not unusual example taken from the colored moving pictures. Suppose a patch of green to appear somewhere on the cinema screen, and furthermore suppose that, perhaps on account of absentmindedness, you have as yet attached no definite meaning to this expanse of green. The green so far means neither green water as viewed in the distance nor the green of foliage; it is simply greenness to you and that is about all. Suddenly you come to realize that this patch of green represents, *means*, the blue sky, and instantly thereupon the green is transformed into a fairly natural appearing area of blue sky. Thereafter, no matter how hard you try to change the blue back into green, you will fail. Here we observe one of the operations of the law of psychological relativity. The traditional psychology taught that a color possesses a more or less absolute, unchanging character. Green, for instance, according to this older mode of thinking, should be green anywhere, regardless of the meaning which it bears. But the simple observation just described would seem to show conclusively that color quality is a matter realtive to the meaning borne by the quality. . . .

The intimate relation between space and time is being brought home to us constantly. Spatial values are shrinking from year to year. A few hours now suffice for the traveler to span the continent. A giant airship flies around the world in a few days. As speed increases, distance decreases. So we come to say that distance depends on, is a function of, time. As in the practical activities of our daily lives we recognize this fundamental fact of relativity, so the psychological laboratory gives detailed information about it.

Take, for instance, such a simple and yet such an enlightening experiment as the following one. Have some person close his eyes, and then, using the blunt point of a pencil, touch him now at one position and then at another on the back of the hand. If you take care to place the two successive touches always the same distance apart as measured in terms of inches, but vary the rate of the touches, you will find that the person upon whom you are experimenting will report that the greater the rate of stimulation the shorter the apparent distance between the touches. And just as in the case of vision it is possible to produce the illusion of motion, so in the field of touch, if the two stimulations are speeded up, there finally comes a stage where the experience of a mere succession of touches is replaced by the experience of one point in continuous motion over the surface of the skin. So it has also been asserted that there is a most curious illusion of movement in hearing, where two sharp sounds are presented first to one ear and then to the other with an extremely brief time interval between the sounds. Throughout all these interesting experiences we have the opportunity of observing the complicated interlocking of space and time. . . .

The explanations of the configuration psychology are couched both in physiological and in physical terms. Take, for instance, the explanation of the illusion of motion advanced by the configurationists. The older and more prevalent account given of the physiological mechanics involved in this illusion has considered absolutely essential the operation of at least two brain levels. We may compare brain levels to the stories of a building, although this analogy has latent in it some very absurd features. The older theory would say that the perception of movement is mediated more particularly by the higher level. The materials entering the lower story from the eyes are in a raw, chaotic state, and it is necessary for the mechanics of the upper story to organize and unify the materials deposited in the lower story before the perception of motion results.

The configuration theory, on the other hand, has speculated on the possibility of the illusion of motion being mediated at the lower brain level, this level doing the entire work. The theory in question invokes a short-circuiting process at the lower brain level, analogous to the short circuit that once in a while occurs in a network of electric wires. The nerves in the brain correspond to the electric wires. This explanation has been set forth by the Gestalt psychologists in terms of the physics of nerve transmission. In fact, as will be pointed out in more detail below, the architecture of the configuration theory emerged out of the abode of modern physics.

III

Primitive man saw in human purpose the expression of the will of good and evil spirits. In nature, the lightning and the whirlwind voiced the wrath of a god. The coming of modern science has enabled us to understand the events of physical nature in terms of natural causes, but the struggle to interpret human action without reference to supernatural agencies has been long drawn out and severe.

Consider the manner in which the configurationist would solve the mysteries of purpose. Cover a section of fine-meshed wire netting with a soap-bubble film. Cast upon this film a loop of thread. In all probability the loop will assume a highly irregular form. Prick the film somewhere within the loop and see what happens. At the moment of pricking, the loop jumps into a shape, a configuration, that shows a strongly circular tendency.

Observing all this, you would not think of saying that the alteration in the form of the loop was brought about by supernatural means. You would seek to explain the fact by reference to surface tensions. Purely physical laws are here active. No genie has directed the shift in form. The little loop of thread has been forced into the shape demanded by the physical law of the expenditure of least possible energy. It has passed into a new state of equilibrium that is in harmony with the change in surrounding conditions, and the tendency has been in the direction of simplification of form, from the irregular to the symmetrical.

The law of least energy, so say the Gestalt psychologists, applies not narrowly to the world of matter but also to the world of mind. Pause for the moment and consider how this law is operative in the fields of perception and memory.

Imagine twelve small white disks against a black background, arranged in such a way that the disks, if joined together by straight lines, would form a regular twelve-sided figure. Now alter this pattern so that any one of the disks is somewhat farther from the center of the figure than the other eleven disks. This alteration has of course destroyed the symmetry of the figure; the pattern now appears to you to be out of balance. If this unbalanced figure is presented for a small fraction of a second a strange thing will be seen to happen. The displaced disk shoots inward toward the position it would occupy were the figure to become a regular, simple, balanced one. Of course, no motion has occurred in the physical sense of that word, but nevertheless the illusion of motion is experienced.

This fact of perception holds many possibilities for a more adequate insight into the nature of mind. Here is to be witnessed the ultimate similarity between the psychological and the physical aspects of the universe. Just as the loop of thread had to pass from a more complex to a simpler form, according to the law of least energy, so did our perception of the unsymmetrical twelve-sided figure undergo a change toward symmetry and simplicity, toward the best possible equilibrium. . . .

The configuration psychology has been a leader in the scientific movement to close the ancient gap between mind and matter. It has interpreted purposeful activity, the seeking after goals, as a natural, not a supernatural, event. To be sure, the configurationists are not the first to have suggested this type of solution. But they have related human and animal purpose more effectively to the facts and theories of modern physics than have their predecessors.

IV

. . . Configurationism and Einstein's doctrine of physical relativity seem to be developing in parallel directions, notwithstanding the fact that the subject-matter of each appears on the surface to be so completely different from that of the other. Einstein protested against the conventional world scheme of Sir Isaac Newton. Einstein begins the story of the physical universe not with such unrealities as absolute space and time but with the account of two bodies in motion relative to each other; he dispenses summarily with a fixed framework of space and time.

For Newton, space and time values are independent of the position of the human observer and are presumed to be identical for every segment of the cosmos. For Einstein, these values depend directly upon the position of the observer. All physical evaluations are purely relative to the particular total situation within which they are calculated. Also, the combination of velocities is not a merely additive matter, but is something more than a straightforward summation. Here we see pronounced the apparently revolutionary proposition that even at the purely physical level the whole is not equal to the sum of the parts.

For the traditional psychology, particular experiences and bits of behavior possess a more or less absolute character. For the Gestalt psychologist, any aspect of mentality has meaning only in its relation to the larger context, the whole, in which it exists. Configurationism begins its narrative with an account of the ordinary meanings of common sense, the wholes of experience, which may be encountered in every-day life. It does not begin with a description of fictitious parts called pure sensations and reflexes.

To the classical psychology, human perception, for instance, is viewed as a mosaic, as a bundle of originally meaningless psychic atoms called simple sensations. For the Gestalt psychology, every perception, whether of a person's face or of anything else, exists in its own right, is itself. The perception is something over and above the sum of its supposed parts.

But the configuration doctrine, in spite of its apparent novelty, owes a big debt to history. What scientific movement does not? Einstein, for that matter, had his predecessors more than two thousand years ago. The configurationists are not the only psychologists to have recognized the idea of psychic relativity, although they are the first to have elaborated the idea into a workable system. The origin of the notion lies in the dim past. Perhaps we might trace its nearer origins to the founder of experimental psychology, Wilhelm Wundt, of the University of Leipzig. But that is another story. Suffice it to say that the concept of relativity in the domain of mind has been evolving slowly but surely. The configurationists, however, merit the distinction of being the first to develop this concept scientifically.

Mind is the outstanding riddle of the universe. Because of the well-nigh unbelievably great complexity of its subject-matter, psychology has lagged in the rear of the conquering advance of the physical sciences.

But, notwithstanding the manifold difficulties confronting the student of mind, it seems certain that the doctrine of psychological relativity as enunciated by the Gestalt school will lead eventually to a thorough readjustment and advancement of psychology at large. Already can there be detected the beneficial influence of this mode of thinking upon current American and European psychology. Whether or not its chief value will finally be found to consist in its rôle as a method rather than as a point of view is a question that need not detain us here.

Whatever the verdict of posterity may be, the Gestalt psychology has attempted the remarkable feat of giving due tribute both to the common-sense meanings that have always been the property of the ordinary scheme of all time, the Einstein doctrine of physical relativity.

9 *José Ortega y Gasset*
"The Historical Significance of the Theory of Einstein"

One of the more perceptive, if sometimes overly acid, critics of the twentieth century was the Spanish philosopher José Ortega y Gasset. It is fitting to close this review of the impact of Relativity Theory by reproducing his discussion of the historical significance of Einstein's theory.

The theory of relativity, the most important intellectual fact that the present time can show, inasmuch as it is a theory, admits of discussion whether it is true or false. But, apart from its truth or falsity, a theory is a collection of thoughts which is born in a mind, in a spirit or in a conscience in the same way as a fruit is born upon a tree. Now, a new fruit indicates that a new vegetable species is making its appearance in the flora of the world. Accordingly, we can study the theory of relativity with the same design as a botanist has in describing a plant: we can put aside the question whether the fruit is beneficial or harmful, whether the theory is true or erroneous, and attend solely to the problem of classifying the new species, the new type of living being which we light upon there. Such an analysis will enable us to discover the historical significance of the theory, viz., its nature as an historical phenomenon.

The peculiarities of the theory of relativity point to certain specific tendencies in the mind which has created it. And as a scientific edifice of this magnitude is not the work of one man but the result of the inadvertent collaboration of many, of all the best contemporary minds, in fact, the orientation which these tendencies reveal will indicate the course of western history.

I do not merely mean by this that the triumph of the theory

SOURCE. José Ortega y Gasset, *The Modern Theme*, James Cleugh, tr., New York: Harper Torchbooks, 1961, pp. 135–145 and 147–152. Reprinted by permission of Harper & Row, Publishers, Inc.

will influence the spirit of mankind by imposing on it the adoption of a definite route. That is an obvious banality. What is really interesting is the inverse proposition: the spirit of man has set out, of its own accord, upon a definite route, and it has therefore been possible for the theory of relativity to be born and to triumph. The more subtle and technical ideas are, the more remote they seem from the ordinary preoccupations of men, the more authentically they denote the profound variations produced in the historical mind of humanity. . . .

1. ABSOLUTISM

. . . Classical mechanics recognises the common relativity of all our conclusions on the question of movement and, therefore, the relativity of every position in space and time which the human mind can observe. How is it, then, that the theory of Einstein which, we are told, has destroyed the entire edifice of classical mechanics, throws into relief in its very name, as its principal characteristic, relativity itself? This is the multiform equivocation which we are bound, above all, to expose. *The relativism of Einstein is strictly inverse to that of Galileo and Newton.* For the latter the empirical conclusions we come to concerning duration, location and movement are relative because they believe in the existence of absolute space, time and movement. We cannot perceive them immediately; at most we possess indirect indications of them (centrifugal forces are an example.) But if their existence is believed in all the effective conclusions we come to will be disqualified as mere appearances, values relative to the standpoint of comparison occupied by the observer. Consequently, relativism here connotes failure. The physical science of Galileo and Newton is relative in this sense.

Let us suppose that, for one reason or another, a man considers it incumbent upon him to deny the existence of those unattainable absolutes in space, time and transference. At once those concrete conclusions, which formerly appeared relative in the sinister sense of the word, being freed from comparison with the absolute, become the only conclusions that express reality. Absolute (unattainable) reality and a further reality, which is relative in comparison with the former, will not now exist. There will only be one single reality, and this will be what positive physics approximately describes. Now, this reality is what the observer perceives from the place he occupies; it is therefore a relative reality. But as this relative reality, in the suppositious case we have taken, is the only one there is, it must, as well as being relative, be true or, what comes to the same thing, absolute reality. Relativism is not here opposed to absolutism; on the contrary, it merges with it and, so far from suggesting a failure in our knowledge, endows the latter with an absolute validity.

This is the case with the mechanics of Einstein. His physical science is not relative, but relativist, and achieves, thanks to its relativism, an absolute significance.

The most absurd misrepresentation which can be applied to the new mechanics is to interpret it as one more offspring of the old philosophic relativism, of which it is in fact the executioner. In the old relativism our knowledge is relative because what we aspire to know, viz., space-time reality, is absolute and we cannot attain to it. In the physics of Einstein our knowledge is absolute; it is reality that is relative.

Consequently, we are above all bound to note as one of the most genuine features of the new theory its absolutist tendency in the sphere of knowledge. It is inexplicable that this point should not have been emphasized as a matter of course by those who interpret the philosophic significance of this innovation of genius. The tendency is perfectly clear, however, in the capital formula of the whole theory: physical laws are true whatever may be the system of reference used, that is to say, whatever the point of observation may be. Fifty years ago thinkers were preoccupied with the question whether "from the point of view of Sirius" human truths would be valid. This is equivalent to a degradation of the science practised by man by an attribution to it of a purely domestic value. The mechanics of Einstein permit our physical laws to harmonise with those which may be conjectured to prevail in minds inhabiting Sirius. . . .

2. PERSPECTIVISM

The provincial spirit has always, and with good reason, been accused of stupidity. Its nature involves an optical illusion. The provincial does not realise that he is looking at the world from a decentralised position. He supposes, on the contrary, that he is at the centre of the whole earth, and accordingly passes judgment on all things as if his vision were directed from that centre. This is the cause of the deplorable complacency which produces such comic effects. All his opinions are falsified as soon as they are formulated because they originate from a pseudo-centre. On the other hand, the dweller in the capital knows that his city, however large it may be, is only one point of the cosmos, a decentralised corner of it. He knows, further, that the world has no centre, and that it is therefore necessary, in all our judgments, to discount the peculiar perspective that reality offers when it is looked at from our own point of view. This is the reason why the provincial always thinks his neighbour of the great city a sceptic, though the fact is that the latter is only better informed.

The theory of Einstein has shown modern science, with its exemplary discipline—the *nuova scienza* of Galileo, the proud physical philosophy of the West—to have been labouring under an acute form of provincialism. Euclidian geometry, which is only applicable to what is close at hand, had been extended to the whole universe. In Germany to-day the system of Euclid is beginning to be called "proximate geometry" in contradistinction to other collections of axioms which, like those of Riemann, are long-range geometries.

The refutation of this provincial geometry, like that of all provincialism, has been accomplished by means of an apparent limitation, an exercise of modesty in the claims of its conqueror. Einstein is convinced that to talk of Space is a kind of megalomania which inevitably introduces error. We are not aware of any more extensions than those we measure, and we cannot measure more than our instruments can deal with. These are our organ of scientific vision; they determine the spatial structure of the world we know. But as every other being desirous of constructing a system of physics from some other place in the earth is in the same case the result is that there is no real limitation involved at all.

There is no question, then, of our relapsing into a subjectivist interpretation of knowledge, according to which the truth is only true for a pre-determined subjective personality. According to the theory of relativity, the event *A*, which from the mundane point of view precedes the event *B* in time, will, from another place in the universe–Sirius, for example–seem to succeed *B*. There cannot be a more complete inversion of reality. Does it mean that either our own imagination or else that of the mind resident in Sirius is at fault? Not at all. Neither the human mind nor that in Sirius alters the conformation of reality. The fact of the matter is that one of the qualities proper to reality is that of possessing perspective, that is, of organising itself in different ways so as to be visible from different points. Space and time are the objective ingredients of physical perspective, and it is natural that they should vary according to the point of view. . . .

When we see a stationary and solitary billiard ball we only perceive its qualities of colour and form. But suppose another ball collides with the first. The latter is then driven forward with a speed proportionate to the shock of the collision. Thereupon we note a new quality of the ball, which was previously latent, viz., its resilience. But, someone may say, resilience is not a quality of the first ball, for the quality in question only appears when the second ball collides with it. We shall answer at once that it is not so. Resilience is a quality of the first ball no less than its colour and form, but it is a reactive quality, i.e., one responsive to the action of another object. Thus, in a man, what we usually call his character is his way of reacting to externality–things, persons or events.

Well, now: when some reality collides with another object which we denominate "conscious subject," the reality responds to the subject by *appearing to it.* Appearance is an objective quality of the real, its response to a subject. This response is, moreover, different according to the condition of the observer; for example, according to his standpoint of contemplation. It is to be noted that perspective and point of view now acquire an objective value, though they were previously considered to be deformations imposed by the subject upon reality. Time and space are once more, in defiance of the Kantian thesis, forms of the real.

If there had been among the infinite number of points of view an exceptional one to which it might have been possible to assign a superior correspondence with nature, we could have considered the rest as deforming agents or as "purely subjective." Galileo and Newton believed that this was the case when they spoke of absolute space, that is to say, of a space contemplated from a point of view which is in no way concrete. Newton calls absolute space *sensorium Die*, the visual organ of God; or, we might say, divine perspective. But we have scarcely thought out in all its implications this idea of a perspective which is not seen from any determined and exclusive place when we discover its contradictory and absurd nature. There is no absolute space because there is no absolute perspective. To be absolute, space has to cease being real—a space full of phenomena—and become an abstraction.

The theory of Einstein is a marvellous proof of the harmonious multiplicity of all possible points of view. If the idea is extended to morals and aesthetics, we shall come to experience history and life in a new way.

The individual who desires to master the maximum amount possible of truth will not now be compelled, as he was for centuries enjoined, to replace his spontaneous point of view with another of an exemplary and standardised character, which used to be called the "vision of things *sub specie aeternitatis.*" The point of view of eternity is blind: it sees nothing and does not exist. Man will henceforth endeavour, instead, to be loyal to the unipersonal imperative which represents his individuality.

It is the same with nations. Instead of regarding non-European cultures as barbarous, we shall now begin to respect them, as methods of confronting the cosmos which are equivalent to our own. There is a Chinese perspective which is fully as justified as the Western.

3. ANTIUTOPIANISM OR ANTIRATIONALISM

The same tendency which in its positive form leads to perspectivism signifies in its negative form hostility to utopianism.

The utopian conception is one which, while believing itself to arise from "nowhere," yet claims to be valid for everyone. To a sensibility of the type evident in the theory of relativity this obstinate refusal to be localised necessarily appears overconfident. There is no spectator of the cosmic spectacle who does not occupy a definite position. To want to see something and not to want to see it from some particular place is an absurdity. Such puerile insubordination to the conditions imposed on us by reality, such incapacity for the cheerful acceptance of destiny, so ingenuous an assumption that it is easy to substitute our sterile desires, are features of a spirit which is to-day nearing its end and on the verge of giving place to another completely antagonistic to it.

The utopian creed has dominated the European mind during the whole of the modern epoch in science, in morals, in religion and in art. The whole weight of the intensely earnest desire to master reality—a specifically European characteristic—had to be thrown into the scales to prevent Western civilisation from perishing in a gigantic fiasco. For the most troublesome feature of utopianism is not that it gives us false solutions to problems—scientific or political—but something worse: the difficulty is that it does not accept the problem of the real as it is presented, but immediately, viz., *a priori*, imposes a form on it which is capricious.

If we compare Western life with that of Asia—Indian or Chinese—we are at once struck by the spiritual instability of the European as opposed to the profound equilibrium of the Oriental mind. This equilibrium reveals the fact that, at any rate in the greatest problems of life, the Easterner has discovered formulae more perfectly adjusted to reality. The European, on the other hand, has been frivolous in his appreciation of the elemental factors of life and has contrived capricious interpretations of them which have periodically to be replaced.

The utopist aberration of human intelligence begins in Greece and occurs wherever rationalism reaches the point of exacerbation. Pure reason constructs an exemplary world—a physical or political cosmos—in the belief that it is the true reality and must therefore supplant the actually existent one. The divergence between phenomena and pure ideas is such that the conflict is inevitable. But the rationalist is sure that the struggle will result in the defeat of reality. *This conviction is the main characteristic of the rationalist temperament.*

Reality, naturally, possesses more than sufficient toughness to resist the assaults of ideas. Rationalism then looks for a way out: it recognises that, *for the moment*, the idea cannot be realised, but believes that success will be achieved in an "infinite process" (Leitbnitz, Kant). . . .

One part of the work of Kant will remain imperishable, viz., his great discovery that experience is not only the aggregate of data transmitted by the senses, but also a product of two factors. The sensible datum has to be received, given its correct affiliation and organised in a system of disposition. This order is supplied by the subjective personality and is *a priori*. In other words, physical experience is a compound of observation and geometry. Geometry is a pentagraph elaborated by pure reason: observation is the work of the senses. All science which is explanatory of material phenomena has contained, contains and will contain these two ingredients.

This identity of composition, invariably exhibited by modern physics throughout its entire history, does not, however, exclude the most profound variations in its spirit. The mutual relation maintained between its two ingredients leaves room, in fact, for very diverse interpretations. Which of the two is to supplant the other? Ought observation to yield to the demands of geometry, or geometry to observation? To decide one way or the other will mean our adherence to one of two antagonistic types of intellectual tendency. There is room for two opposed castes of opinion in one and the same system of physics.

It is common knowledge that the experiment of Michelson is crucial in the hierarchy of such tests: physical theory is there placed between the devil and the deep sea. The geometrical law which proclaims the unalterable homogeneity of space, whatever may be the processes which occur in it, enters into uncompromising conflict with observation, in fact, with matter. One of two things must happen: either matter is to yield to geometry or the latter to the former.

In this acute dilemma two intellectual temperaments come before us, and we are able to observe their reaction. Lorentz and Einstein, confronted by the same experiment, take opposite resolutions. Lorentz, in this particular representing the old rationalism, believes himself obliged to conclude that it is matter which yields and contracts. The celebrated "contradiction of

Lorentz" is an admirable example of utopianism. It is the Oath of the Tennis Court transferred to physics. Einstein adopts the contrary solution. Geometry must yield, pure space is to bow to observation, to curve, in fact.

In the political sphere, supposing the analogy to be a perfect one, Lorentz would say: Nations may perish, provided we keep our principles. Einstein, on the other hand, would maintain: We must look for such principles as will preserve nations, because that is what principles are for.

It is not easy to exaggerate the importance of the change of course imposed by Einstein upon physical science. Hitherto the *rôle* of geometry, of pure reason, has been to exercise an undisputed dictatorship. Common speech retains a trace of the sublime function which used to be attributed to reason: people talk of the "dictates of reason." For Einstein the *rôle* of reason is a much more modest one: it descends from dictatorship to the status of a humble instrument, which has, in every case, first to prove its efficiency.

Galileo and Newton made the universe euclidian simply because reason dictated it so. But pure reason cannot do anything but invent systems of methodical arrangement. These may be very numerous and various. Euclidian geometry is one, Riemann's another, Lobatchewski's another, and so on. But it is clearly not these systems, not pure reason, which resolve the nature of the real. On the contrary, reality selects from among these possible orders or schemes the one which has most affinity with itself. This is what the theory of relativity means. The rationalist past of four centuries is confronted by the genius of Einstein, who inverts the time-honoured relation which used to exist between reason and observation. Reason ceases to be an imperative standard and is converted into an arsenal of instruments; observation tests these and decides which is the most convenient to use. The result is the creation of the science of mutual selection between pure ideas and pure facts.

This is one of the features which it is most important to emphasize in the thought of Einstein, for here we discover the initiation of an entirely new attitude to life. Culture ceases to be, as hitherto, an imperative standard to which our existence has to conform. We can now see a more delicate and more just relation between the two factors. Certain phenomena of life are selected as possible forms of culture; but of these possible forms of culture life, in its turn, selects the only ones which are suitable for future realisation.

4. FINITISM

I should not like to conclude this genealogical sketch of the profound tendencies rife in the theory of relativity without alluding to the most clear and patent of them. While the utopist past used to settle all disputes by the expedient of recourse to the infinite in space and time, the physics of Einstein . . . annotates the universe. The world of Einstein is curved, and therefore closed and finite. . . .

This gives an enormous range of reference to the fact that physics and mathematics are suddenly beginning to have a marked preference for the finite and a great distaste for the infinite. Can there be a more radical difference between two minds than that one should tend to the idea that the universe is unlimited and that the other should feel its environment to be circumscribed? The infinity of the cosmos was one of the great intoxicating ideas produced by the Renaissance. It flooded the hearts of men with tides of pathetic emotion, and Giordano Bruno suffered a cruel death on its behalf. During the whole of the modern epoch the most earnest desires of Western man have concealed, as though it were a magical foundation for them, this idea of the infinity of the cosmic scene.

And now, all at once, the world has become limited, a garden surrounded by confining walls, an apartment, an interior. Does not this new setting suggest an entirely different style of living, altogether opposed to that at present in use? Our grandsons will enter existence armed with this notion, and their attitude to space will have a meaning contrary to that of our own. There is evident in this propensity to finitism a definite urge towards limitation, towards beauty of serene type, towards antipathy to vague superlatives, towards antiromanticism. The Greek, the "classical" man, also lived in a limited universe. All Greek culture has a horror of the infinite and seeks the *metron*, the mean.

It would be superficial, however, to believe that the human mind is being directed towards a new classicism. There has never yet been a new classicism which has not resulted in frivolity. The classical man seeks the limit, but it is because he has never lived in an unlimited world. Our case is inverse: the limit signifies an amputation for us, and the closed and finite world in which we are now to draw breath will be, irremediably, a truncated universe.

A SELECTED BIBLIOGRAPHY

The standard biography of Sir Isaac Newton is L. T. More, *Isaac Newton*, New York and London, 1934. It is rapidly being rendered obsolete by modern research. The student should also consult E. C. da C. Andrade, *Isaac Newton*, New York, 1950, and A. Koyré, *Newtonian Studies*, London, 1965. For Ernst Mach's ideas see Ernst Mach, *The Analysis of Sensations and the Relation of the Physical to the Psychical*, Chicago and London, 1914, and *Space and Geometry*, Lasalle, Ill., 1906. There is no good account of Mach's thought in English. Robert Bouvier's *La pensée d'Ernst Mach*, Paris, 1923, and Hugo Dingler's *Die Grundgedanken der Machschen Philosophie*, Leipzig, 1924, discuss different aspects of Mach's philosophy. Henri Poincaré's popular works on the philosophy of science were collected together and published in English under the title *The Foundations of Science*, New York and Garrison, N. Y., 1913. They are well worth reading both for the light they shed on Poincaré's ideas and for their general illumination of the problem of the nature of scientific thought. H. A. Lorentz is probably the most neglected great modern scientist. For a very short, and totally inadequate, account of his achievement, see G. L. de Haas-Lorentz, editor, *H. A. Lorentz, Impressions of His Life and Work*, Amsterdam, 1957.

For general accounts of the history of physical thought from Newton to Einstein, see, A. d'Abro, *The Evolution of Scientific Thought from Newton to Einstein*, New York, 1950, a difficult, and not always trustworthy, historical narrative. See also Hans Reichenbach, *From Copernicus to Einstein*, New York, 1942.

The number of books on Relativity Theory is immense. Einstein himself was an eager popularizer of his own ideas. Besides his book on Relativity cited in the text, see A. Einstein and L. Infeld, *The Evolution of Physics*, New York, 1938, an excellent and simple explanation of the theory of Special and General Relativity. One of the most popular and most readable accounts of Relativity Theory is that by Lincoln K. Barnett, *The Universe and Dr. Einstein,* New York, 1948. A bit more difficult is Hermann Bondi, *Relativity and Common Sense; a new approach to Einstein,* Garden City, N. Y., 1964. Max Born's, *Einstein's Theory of Relativity*, London, 1924, and P. W. Bridgman, *A Sophisticate's Primer of the Special Theory of Relativity*, Middletown, Conn., 1962, require some knowledge of physics and mathematics. G. J. Whitrow, *The Natural Philosophy of Time*, London, 1961, is an excellent survey of the history of the notion of time, including the revolution created by Einstein.

The problem of the ether after Einstein's 1905 paper is discussed in Oliver Lodge, *The Ether of Space*, New York, 1909, and *Ether and Reality*, New York, 1925. Einstein's views are to be found in his article, "Ether and the Theory of Relativity," in A. Einstein, *Sidelights on Relativity*, London, 1922.

There is no first-rate biography of Einstein. For his life, consult Philipp Frank, *Einstein, His Life and Times*, New York, 1947, and Leopold Infeld, *Albert Einstein, His Work and Its Influence on Our World*, New York, 1950.

The impact of Relativity Theory on Philosophy has been well described in Milic Capek's *The Philosophical Impact of Contemporary Physics*, New York, 1961. Adolf Grünbaum has analyzed the *Philosophical Problems of Space and Time* in a very difficult book published in New York in 1963. The reader of this work must know his physics and his philosophy of science well. It is the best work on the subject and will repay serious study.

For other philosophical reactions to Relativity Theory, see H. Weyl, *Space-Time-Matter*, New York, 1950; A. S. Eddington, *The Nature of the Physical World*, New York, 1928; and his *Space, Time and Gravitation*, Cambridge, 1953. See also Hans Reichenbach, *The Philosophy of Space and Time*, New York, 1958, and *The Theory of Relativity and a priori Knowledge*, Berkeley, California, 1965; Henri Bergson, *Duration and Simultaneity, with reference to Einstein's Theory*, Indianapolis, 1965; and Ernst Cassirer, *Substance and Function, and Einstein's Theory of Relativity*, Chicago, 1923.

The history of the impact of Relativity Theory on modern thought remains to be written. For some clues to this impact, see G. D. Birkhoff, *The Origin, Nature and Influence of Relativity*, New York, 1925.